Source Water Protection for Concentrated Animal Feeding Operations:

A Guide for Drinking Water Utilities

About the Awwa Research Foundation

The Awwa Research Foundation (AwwaRF) is a member-supported, international, nonprofit organization that sponsors research to enable water utilities, public health agencies, and other professionals to provide safe and affordable drinking water to consumers.

The Foundation's mission is to advance the science of water to improve the quality of life. To achieve this mission, the Foundation sponsors studies on all aspects of drinking water, including supply and resources, treatment, monitoring and analysis, distribution, management, and health effects. Funding for research is provided primarily by subscription payments from approximately 1,000 utilities, consulting firms, and manufacturers in North America and abroad. Additional funding comes from collaborative partnerships with other national and international organizations, allowing for resources to be leveraged, expertise to be shared, and broad-based knowledge to be developed and disseminated. Government funding serves as a third source of research dollars.

From its headquarters in Denver, Colorado, the Foundation's staff directs and supports the efforts of more than 800 volunteers who serve on the board of trustees and various committees. These volunteers represent many facets of the water industry, and contribute their expertise to select and monitor research studies that benefit the entire drinking water community.

The results of research are disseminated through a number of channels, including reports, the Web site, conferences, and periodicals.

For subscribers, the Foundation serves as a cooperative program in which water suppliers unite to pool their resources. By applying Foundation research findings, these water suppliers can save substantial costs and stay on the leading edge of drinking water science and technology. Since its inception, AwwaRF has supplied the water community with more than $300 million in applied research.

More information about the Foundation and how to become a subscriber is available on the Web at **www.awwarf.org.**

Source Water Protection for Concentrated Animal Feeding Operations:

A Guide for Drinking Water Utilities

Prepared by:
Richard W. Gullick, **Richard A. Brown**, and **David A. Cornwell**
Environmental Engineering & Technology, Inc.
712 Gum Rock Court, Newport News, VA 23606

Jointly sponsored by:
Awwa Research Foundation
6666 West Quincy Avenue, Denver, CO 80235-3098
and
U.S. Environmental Protection Agency
Washington D.C.

Published by:

Awwa Research Foundation

American Water Works Association

IWA Publishing

DISCLAIMER

This study was jointly funded by the Awwa Research Foundation (AwwaRF) and the U.S. Environmental Protection Agency (USEPA) under Cooperative Agreement No. CR-83110401. AwwaRF and USEPA assume no responsibility for the content of the research study reported in this publication or for the opinions or statements of fact expressed in the report. The mention of trade names for commercial products does not represent or imply the approval or endorsement of AwwaRF or USEPA. This report is presented solely for informational purposes.

CONTENTS

TABLES

FIGURES

FOREWORD

The Awwa Research Foundation (AwwaRF) is a nonprofit corporation that is dedicated to the implementation of a research effort to help utilities respond to regulatory requirements and traditional high-priority concerns of the industry. The research agenda is developed through a process of consultation with subscribers and drinking water professionals. Under the umbrella of a Strategic Research Plan, the Research Advisory Council prioritizes the suggested projects based upon current and future needs, applicability, and past work; the recommendations are forwarded to the Board of Trustees for final selection. The foundation also sponsors research projects through the unsolicited proposal process; the Collaborative Research, Research Applications, and Tailored Collaboration programs; and various joint research efforts with organizations such as the U.S. Environmental Protection Agency, the U.S. Bureau of Reclamation, and the Association of California Water Agencies.

This publication is a result of one of these sponsored studies, and it is hoped that its findings will be applied in communities throughout the world. The following report serves not only as a means of communicating the results of the water industry's centralized research program but also as a tool to enlist the further support of the nonmember utilities and individuals. Projects are managed closely from their inception to the final report by the foundation's staff and large cadre of volunteers who willingly contribute their time and expertise. The foundation serves a planning and management function and awards contracts to other institutions such as water utilities, universities, and engineering firms. The funding for this research effort comes primarily from the Subscription Program, through which water utilities subscribe to the research program and make an annual payment proportionate to the volume of water they deliver and consultants and manufacturers subscribe based on their annual billings. The program offers a cost-effective and fair method for funding research in the public interest.

A broad spectrum of water supply issues is addressed by the foundation's research agenda: resources, treatment and operations, distribution and storage, water quality and analysis, toxicology, economics, and management. The ultimate purpose of the coordinated effort is to assist water suppliers to provide the highest possible quality of water economically and reliably. The true benefits are realized when the results are implemented at the utility level. The foundation's trustees are pleased to offer this publication as a contribution toward that end.

Walter J. Bishop
Chair, Board of Trustees
Awwa Research Foundation

Robert C. Renner, P.E.
Executive Director
Awwa Research Foundation

ACKNOWLEDGMENTS

This research project and preparation of this report was funded by the Awwa Research Foundation (AwwaRF). The project team gratefully acknowledges the assistance of the AwwaRF project managers, Mr. Ryan Ulrich and Ms. Linda Reekie. The project team is also thankful for the assistance, guidance and valuable critiques provided by the AwwaRF Project Advisory Committee:

- Mr. Kim Coy (Watershed Superintendent, City of Akron - Water Supply Division; Akron, Ohio)
- Ms. Wyndi Klement (Senior Research Analyst, Advanced Concepts and Technologies International, LLC; Waco, Texas)
- Mr. Jeff Jollie (Hydrogeologist, US Environmental Protection Agency, Office of Ground Water and Drinking Water; Washington, DC)

Instrumental to the project were the team of expert technical advisors who provided guidance for information resources and a critical review of the potential strategies utilities could employ. The technical advisors included the following:

- Dr. Lori Marsh (Associate Professor, Virginia Polytechnic Institute and State University; Blacksburg, Virginia). Dr. Marsh also wrote part of the sections on poultry, swine and aquaculture operations in Chapter 4
- Mr. Curt Gooch (Senior Extension Associate, College of Agriculture and Life Sciences, Cornell University; Ithaca, New York)
- Dr. Brent W. Auvermann (Associate Professor, Texas A&M University, Agricultural Research and Extension Center at Amarillo, Texas)
- Dr. Thomas Harter (Groundwater Hydrology Specialist (Cooperative Extension), University of California at Davis, Department of Land, Air, and Water Resources; Davis, California)
- Mr. Peter Wright (NY State Conservation Engineer, USDA-NRCS; Syracuse, New York)
- Mr. Victor Johnson (Technical Service Manager - Agricultural and Environmental Services; Kemiron Corp.; Phil Campbell, Alabama)

In addition, Thierry Sam Tamers and Troy Scott of Source Molecular Corporation (Miami, Florida) wrote much of the section on microbial source tracking in Chapter 6.

The input from a wide variety of drinking water utilities is also gratefully acknowledged. Sixty four (64) US water utilities completed the project survey questionnaire. In addition, the authors would like to thank the following participating utilities:

- Iowa City Water Division (Iowa City, Iowa)
- Mankato Municipal Water Supply (Mankato, Minnesota)
- Minneapolis Water Works (Minneapolis, Minnesota)
- Saint Cloud Public Utilities (Saint Cloud, Minnesota)
- Saint Paul Regional Water Services (Saint Paul, Minnesota)
- Kansas City Water Services Dept. (Kansas City, Missouri)
- Lincoln Water System (Lincoln, Nebraska)

- New York City Dept. of Environmental Protection (Flushing, New York)
- City of Syracuse Department of Water (Syracuse, New York)
- Wilmington Water & Sewer (Wilmington, North Carolina)
- City of Columbus DOW (Columbus, Ohio)
- Tulsa Metropolitan Utility Authority (Tulsa, Oklahoma)
- Philadelphia Water Department (Philadelphia, Pennsylvania)
- Amarillo Municipal Water System (Amarillo, Texas)
- Weber Basin Water Conservancy District (Layton, Utah)
- City of Cedar Rapids Water Department (Cedar Rapids, Iowa)

The assistance and cooperation of the following organizations is also greatly appreciated:

- Delaware County Extension Office, Oklahoma Cooperative Extension Service (Jay, Oklahoma)
- North Carolina Department of Environment and Natural Resources, Division of Water Quality (Wilmington, North Carolina)
- Onondaga County Soil and Water Conservation District (LaFayette, New York)
- Texas Cattle Feeders Association (Amarillo, Texas)

Dr. Christophe Darnault was instrumental in the proposal preparation and early planning stages of this project before he resigned from EE&T, Inc. to work at the University of Illinois-Chicago. Mr. Ebert Obando (EE&T) provided assistance during compilation and evaluation of information collected during this study. Mr. William Cornwell (EE&T) created programming of the web-based information center accompanying this report.

EXECUTIVE SUMMARY

Animal feeding operations (AFOs) for beef cattle, dairy cattle, swine, chickens, turkeys, and many other species are located throughout the United States to meet the demand for meat and other animal-related products. Byproducts of animal production include manure (feces and urine), animal parts, animal mortality, feed lot runoff, excess feed, and others. The growing trend towards large farms that concentrate increasing numbers of animals into smaller areas has magnified waste handling difficulties and increased the potential for releases of excess animal feed, waste, or byproducts into the environment. The larger of these farms are defined by the US Environmental Protection Agency (USEPA) as concentrated animal feeding operations (CAFOs). USEPA claims that improperly managed manure from these operations "has caused serious acute and chronic water quality problems" (USEPA 2003a).

The USEPA promulgated new regulations for AFOs and CAFOs (the 'CAFO Rule'; USEPA 2003a) that became effective on April 14, 2003. The CAFO Rule is intended to simplify and clarify the existing regulations, and requires CAFOs to: (a) be permitted as a point source; (b) develop plans for storing, using, and disposing of animal wastes; and (c) to prepare annual compliance reports on waste disposal practices. The specific size criteria that determine whether an animal farm is a CAFO or AFO are presented in Table 4.2 of this report.

The information and guidance presented in this report will help drinking water utilities better understand the basics of animal feeding operations and related agricultural activities, the characteristics and quantity of the wastes produced, the fate and transport of these contaminants in the environment, potential contaminant impacts on source water quality, and strategies that water utilities and CAFOs may use to control and monitor the release of these contaminants to drinking water sources. The information was obtained from a variety of sources including review of available literature, consultation with utility participants and a team of expert technical advisors, a survey of drinking water utilities, and in-depth examination of related source water protection case studies, including site visits to a variety of water utilities and CAFOs in Texas, Oklahoma, Arkansas, North Carolina, and New York. As a complement to this report, an electronic information center (www.eetinc.com/cafo) was developed that can further help a utility to evaluate their specific situation and identify potential cost-effective proactive measures.

ANIMAL FEEDING OPERATION IMPACTS ON SOURCE WATER QUALITY

Drinking water utilities may be adversely affected by impairments in source water quality due to contaminants originating at AFOs. Source water contamination from AFO wastes can occur because of normal operations and substandard practices, nutrient-excessive land application of manure, and major incidents such as breaches of large liquid manure storage facilities.

For surface waters, the most common problems for water suppliers associated with AFO wastes in their source water area are excessive nutrients (primarily phosphorus and nitrogen) causing eutrophication and resulting increases in organic matter and taste and odor, as well as the potential presence of microbial pathogens such as *Cryptosporidium*. For groundwater supplies, nitrate and pathogens have generally caused the most problems (USEPA 2003a). Other impacts related to AFOs can include salts, organic matter, solids, and volatile and odorous compounds. To a lesser extent, animal wastes may also be a source of trace elements, antibiotics, hormones, and pesticides. Furthermore, AFOs may also have waste containing soaps, surfactants, acids, alkalines, and disinfectants used for cleaning, health maintenance, and food safety.

Animal waste constituents can be released to the environment during any of the waste generation and handling steps, including the animal confinement areas, treatment and storage lagoons, manure stockpiles, and land application processes. The species, size, maturity and health of the animals, as well as the composition of their feed, affect the characteristics of the manure and other wastes. The main transport pathways include surface runoff and erosion, overflows from liquid waste storage facilities (e.g., lagoons and storage ponds), spills and other dry-weather discharges, leaching into soil and ground water, and volatilization of compounds (e.g., ammonia) and subsequent redeposition on land or surface water (USEPA 2003a). Preferential flow paths can be especially important pathways for subsurface transport of contaminants.

REGULATION OF ANIMAL FEEDING OPERATIONS

The largest of these animal production facilities that meet definitions outlined in USEPA regulations are called Concentrated Animal Feeding Operations (CAFOs). Starting in April 2003, a revised federal regulation (USEPA 2003a) took effect requiring CAFOs to: (a) be permitted as a point source under the National Pollutant Discharge Elimination System (NPDES); (b) develop plans for storing, using, and disposing of animal wastes; and (c) prepare annual compliance reports on waste disposal practices.

The USEPA estimates there are over 15,500 CAFOs in the US that manage about 300 million tons of manure annually (USEPA 2003a). All subject CAFOs must develop and implement a site-specific Comprehensive Nutrient Management Plan (CNMP), including a plan for land application when this option is used. The USEPA and USDA also recommend and encourage all other AFOs to use a site-specific, economically feasible, and technically sound CNMP, and to participate in voluntary programs conducted by federal, state, and local agencies to learn about technical and financial assistance available for developing or implementing a CNMP for their operation.

Other laws may also be applicable to CAFO operations, including existing state laws, individual state CAFO requirements beyond the USEPA CAFO Rule regulations, Total Maximum Daily Load (TMDL) programs implemented as part of the Clean Water Act, Coastal Zone Management Reauthorization, and others.

BEST MANAGEMENT PRACTICES FOR ANIMAL FEEDING OPERATIONS

All farms and AFOs within a source water area should have farm management plans that include descriptions of appropriate waste collection and transfer techniques designed to minimize the release of contaminants to the environment. Planning should include a pollutant mass balance analysis to assess if the farm is a significant contributor of pollutants to the environment. Best management practices (BMPs) for animal feeding operations include a variety of proactive solutions that reduce the amount of contaminant waste created, as well as reactive solutions for managing and treating the wastes produced.

Waste minimization techniques include dietary management (feed type optimization), animal health maintenance, clean water diversion and stormwater run-on control, animal watering control practices, fencing to keep animals out of streams, soil erosion prevention, and appropriate tilling, planting, and irrigation practices. Reactive solutions for managing and treating wastes include stormwater runoff control and treatment (including vegetative filter strips), replacing earthen barnyards with covered shelters or concrete pads with curbs (to facilitate cleaning and direct polluted water to storage or treatment), finding alternative uses for manure (e.g., odor-free fertilizers, soil amendments, feed additives, and energy production), appropriate land application of manure at agronomic rates that do not adversely affect source water quality, and various treatment practices.

Many different types of physical, chemical and microbiological treatment systems are available to manage AFO wastes. For example, solid-liquid separation, composting, lagoons, and anaerobic digesters are used to treat manure. Alum residuals from water treatment plants may be used to bind phosphorus and reduce its solubilization after land application of manure and litter (USEPA 2004b). Common techniques for disposal of dead animals include composting, rendering, incineration, burial, and landfilling. Spill prevention and control measures should be used for stored fuels and chemicals, and emergency action plans prepared for potential major events such as large-scale spills of liquid manure, lagoon water, fuel oils, and other liquid materials.

WATER UTILITY APPROACHES TO SOURCE WATER PROTECTION

Strategic measures that utilities employ to protect source water include source water assessments, source water and early warning monitoring, chemical and microbial contaminant source tracking, source water protection activities, cooperation with stakeholders, outreach and education programs, obtaining outside grant funding, contributing funds for development and maintenance of on-farm BMPs, intake management, pre-sedimentation, off-stream storage, riverbank filtration, and various water treatment processes. Case studies that have successfully used these approaches are described in the report.

Water utilities need to assess what potential waste sources have the greatest potential adverse impact on source water quality, and prioritize protection activities and measures for action in their source water area. While the assessment should identify the type and location of all CAFOs in a source water area, assessment and consideration for prioritization should include all potential waste sources, not just those from the agricultural sector or CAFO facilities alone. As for impacts from the agricultural sector, water utilities should be concerned not just about AFOs and CAFOs, but about poorly managed farms of any type. Periodic reassessment of the action priorities and the success of approaches taken will help ensure the program is achieving the optimal benefits.

Identification of the key pollutant sources through water quality monitoring and source water assessments allows utilities to appropriately prioritize their actions toward the most important potential contamination sources. For example, microbial source monitoring techniques can accurately assess the sources of pathogenic contamination in water supplies, allowing the utility to determine the specific AFO(s) or other pollutant source(s) that are contributing to the water quality problems or risks experienced. Water quality monitoring can also be used to verify effectiveness of implemented BMPs. Early warning systems can provide timely information about rapid changes in water quality so that intake management and treatment practices can be manipulated as necessary.

Monitoring of source water quality is important for identifying pollutants and their sources (to target protection efforts), and for verifying the effectiveness of source water protection programs (including monitoring the effectiveness of BMPs at AFOs). Many monitoring programs target both dry and wet (storm event) sampling of their source water and upstream locations, as the primary pollutant source may vary depending on the amount and recent frequency of rainfall and runoff, the amount of fixed discharges (e.g., public sewage treatment plants), and other factors.

Monitoring programs are used as a primary intake water protection measure by many utilities that have no other means of controlling outside activities in their source water area, including utilities drawing source water from major rivers with extensive watersheds. Perhaps the most important step for source water protection in this case is to develop a dialogue and relationship with the AFOs and CAFOs of concern. Establishing good communication and collaboration between the water utility, farmers, regulators, and other key stakeholders is essential for effective source water protection, and can not be overemphasized. Many AFO operators want to be "good neighbors" and certainly have no desire to pollute the watershed. Such proactive AFOs may be willing to work with regulators and drinking water utilities to develop mutually beneficial partnerships. Working in harmony, reasonable communication can be established and sound, cost-effective control measures put in place.

AFO operators will be more willing to cooperate if they can be shown through sound science that their operations and actions do have an adverse impact on a water source, and that if they take the appropriate corrective measures, reasonable and measurable improvement in source water quality will result. Data and sound science showing agricultural impacts on source water quality should be locally and regionally based in order to be more convincing to local and regional agricultural interests. However, utilities, for their part, will also need to demonstrate understanding of the AFO operators' concerns. Some utilities may prefer to have agriculture in their source water area as opposed to suburban or urban land uses.

One of the most effective means for concerned water utilities to approach agricultural interests is through cooperation with entities that already have relationships established with the local farmers, such as county soil and water conservation districts, cooperative extension offices, US Department of Agriculture (USDA) Natural Resources Conservation Service (NRCS) offices, and other related organizations. Working through local agricultural and watershed stakeholder organizations and coalitions generally builds trust better than when utilities initially approach farms and the agricultural community on their own. State and federal agencies may also be willing to provide financial and technical assistance to help improve water quality. Water utilities should take advantage of these resources as a more effective and more economical approach than trying to do all the work on their own.

Funding is a common obstacle to improvement of agricultural operations for protecting drinking water source quality. Many successful case studies have shown some type of cost-sharing to be effective. Of course, farmer cooperation is best when funding sources are available to pay for 100 percent of the costs of BMP construction and implementation. Proactive water utilities can work with farmers and other stakeholders to find funding for a variety of agricultural source water protection projects. Knowledge of these potential funding sources, as well as good relationships with stakeholder organizations that are experienced at gaining related funding, can greatly help a water utility to gain cooperation of the farms in their source water area.

One excellent measure that water utilities withdrawing surface water in agricultural areas may want to consider is riverbank filtration (RBF), where water is withdrawn from wells located near surface water sources (where most of the groundwater originates from the surface water). RBF essentially uses the subsurface media between the source and groundwater well to remove a variety of contaminants. RBF can reliably remove microbial pathogens, organic contaminants (including taste and odor causing substances and disinfection byproduct (DBP) precursors), hormones, endocrine disruptors, turbidity, and other substances. RBF also dampens or eliminates contaminant spikes.

Other treatment alternatives are also available, including: ultraviolet (UV) disinfection, membranes, alternative oxidants, and other specialized treatment (e.g., arsenic removal). Regulatory pressure requiring compliance from the USEPA Long Term 2 Enhanced Surface

Water Treatment Rule (LT2ESWTR) and Stage 2 Disinfectants and Disinfection Byproducts Rule (DBPR), for example, is influencing utilities to look more closely at these technologies. Several utilities have already installed activated carbon adsorption, UV disinfection, and advanced oxidation or membrane filtration at least in part for treatment of contaminants originating from AFO operations.

As a last resort, some utilities have felt it necessary to take legal action against animal producers that were polluting source waters. Good, sound scientific assessments of source area pollutant loads and impacts are essential for presenting a valid case to force producers to implement pollution prevention measures. As an example, supported by such assessments, the City of Tulsa, Oklahoma was successful in a legal challenge against poultry producers that resulted in substantial reductions in the amount of chicken litter spread on land in the watershed.

CHAPTER 1
INTRODUCTION

Animal feeding operations (AFOs) for beef cattle, dairy cattle, swine, chickens, turkeys, and many other species are located throughout the United States to meet the demand for meat and other animal-related products. The growing trend towards concentrating increasing numbers of animals into smaller areas has magnified waste handling difficulties and increased the potential for releases of excess animal feed, waste, or byproducts into the environment. The amount of waste produced at these facilities can be substantial, and large animal farms can produce as much excreted waste as small and medium-size cities. For example, the US Environmental Protection Agency (USEPA) estimates that a farm with 2,500 dairy cattle is similar in waste load to a city of 411,000 people (USEPA 2004b).

The largest animal feeding operations that meet definitions outlined in USEPA regulations are called Concentrated Animal Feeding Operations (CAFOs). CAFOs are classified as agriculture facilities that house or feed large numbers of animals in a confined area for 45 days or more during any 12-month period. The USEPA estimates there are over 15,500 CAFOs in the US that manage about 300- million tons of manure annually (USEPA 2003a). While only a small fraction of the approximately 257,000 AFOs in the US (USEPA 2003b) can be classified as CAFOs, CAFOs still represent nearly 60 percent of the 500 million tons of manure produced annually by operations that confine animals.

Figure 1.1 shows the number of large animal farms in each state (with greater than 300 animal units, and greater than 1,000 units; the term 'animal units' is defined in Figure 1.1 and is an approach used to standardize farm size across the various animal types). A greater number of large animal farms (based on animal units) are found in Iowa and North Carolina than in the other states (followed by Georgia and California). While both Iowa and North Carolina are major agricultural producers, this ranking is also likely a result of both states having large numbers of farms for two major different species (hogs and chickens). Iowa is ranked #1 in the US in both hog production and chicken egg production, while North Carolina is ranked #2 in hog production and #3 in broiler chicken production. Regional trends are discussed in the sections below for each major animal type.

Surface water quality in the United States has in many places improved greatly since enactment of the Clean Water Act. However, nearly 40 percent of the country's assessed waters show impairments from a range of pollutant sources, and improper management of manure from CAFOs is among the many contributors. USEPA considers that improperly managed manure "has caused serious acute and chronic water quality problems" (USEPA 2003a).

Wastes from AFOs can include manure (feces and urine), spent bedding materials (litter), animal parts, dead animals, and feed lot runoff. The primary pollutants associated with animal wastes are nutrients (primarily nitrogen and phosphorus), pathogens, salts, organic matter, solids, and volatile and odorous compounds. To a lesser extent, animal wastes may also be a source of trace elements, antibiotics, hormones, and pesticides. The species, size, maturity and health of the animals, as well as the composition of their feed, affect the characteristics of the manure.

Number of AFOs (>1,000 AU) / number of AFOs (>300 AU)

◈ = No data reported.

Animal unit (AU) is a unit of measurement for any animal feeding operation calculated by the following formula: AU = (1.0 x number of slaughter and feeder cattle and dairy heifers) + (1.4 x the number of mature dairy cattle) + (0.4 x the number of swine weighing over 55 pounds) + (0.1 x the number of weaned swine weighing 55 pounds or less) + (0.1 x the number of sheep) + (2.0 x the number of horses/mules) + (0.01 x the number of chickens) + (0.02 x the number of turkeys) + 0.2 x the number of ducks). For immature dairy cattle or species not listed, the number of animal units shall be calculated as the average weight of the animals, divided by 1,000 pounds, multiplied by the number of animals.

Source: Data from Tetra Tech, Inc. (2002). Available at www.epa.gov/npdes/pubs/statecom.pdf.

Figure 1.1 Estimated number of large animal farms throughout the US

Source water contamination from AFO wastes can occur because of normal operations and substandard practices, nutrient-excessive land application of manure, and also major acute impacts from large incidents such as breaches of liquid waste storage ponds or lagoons. The concentration of numerous CAFOs in particular regions of the country has resulted in amounts of manure being generated that well exceed the nutrient assimilation capacities of the crop and pasture land that is available for disposal in the nearby area. As such, continued land application has resulted in excessive loads of nutrients and other waste contaminants entering source waters in some watersheds. The number of US counties with manure surpluses more than doubled from 1982 to 1997 (Kellogg et al. 2000). Geographic consolidation of the swine and poultry industry has been particularly pronounced since the 1990s. Portions of Arkansas, Iowa, Nebraska, North Carolina, Georgia, Alabama, and the Delaware-Maryland-Virginia peninsula have AFO manure production well above local crop nutrient requirements, with resulting source water quality issues such as high phosphorus loadings causing eutrophication in surface waters, and levels of nitrate in groundwater exceeding the USEPA Maximum Contaminant Level (MCL) of 10 mg/L.

The number and type of animals determines whether a facility is designated as a CAFO. Federal National Pollutant Discharge Elimination System (NPDES) regulations require CAFOs to be permitted, including development of approved plans to keep contaminants from these operations from impacting groundwater and surface water. Between the 1970s and early 2003, CAFO waste storage and disposal guidelines were voluntary at the federal level, though some states had regulatory programs. Starting in April 2003, a revised USEPA rule for CAFOs (USEPA 2003a; 40 CFR Parts 9, 122, 123, and 412) took effect requiring CAFOs to: (a) be permitted as a point source under the NPDES; (b) develop plans for storing, using, and disposing of animal wastes; and (c) prepare annual compliance reports on waste disposal practices. In some states, CAFOs that have successfully demonstrated no potential to discharge may avoid a permit (USEPA 2003a).

Aquatic animal production facilities (aquaculture) are not subject to the CAFO Rule, and instead are regulated as an NPDES point source via 40 CFR Part 451 (Concentrated Aquatic Animal Production Point Source Category). Both terrestrial and aquatic animal production facilities produce contaminants of concern to drinking water quality, and for purposes of this report the term "AFO-derived" will refer to contaminants from either terrestrial or aquatic animal production facilities.

The trend towards increasing size of feeding operations, coupled with increased sensitivities and stresses on water quality from other contaminant sources, results in a need to focus on the impact of AFOs and CAFOs on source water quality, both now and in the future. The consequences of these activities to the environment in general and humans in particular are poorly understood, even though the potential degradation of the environment, including drinking water sources, is potentially extensive. Information that is available is sometimes not well documented, available in widely-dispersed sources, or unknown or infrequently investigated by water industry practitioners. The Awwa Research Foundation (AwwaRF) has initiated studies to address agricultural and AFO impacts on water supplies, and to provide utilities with tools and guidance to deal with this developing issue (e.g., Fletcher, Davis, and Pyke 2004; Pyke et al. 2003).

Not all utilities have CAFOs in their source water areas, and not all that do experience an adverse effect. However, some utilities have had significant problems for source water quality resulting from CAFO wastes. Many utilities have found they have little input as to whether a CAFO is located in their source water area or what requirements are included in the facility's discharge permit. In fact, they often are not informed of the presence of the CAFO or contents of the permit requirements until after a spill or release takes place.

This report is designed to provide information to utilities to help them be more proactive in protecting their source water area. It includes descriptions of pollution prevention measures for animal feeding and related agricultural operations, discussion of strategies for monitoring and controlling AFO contaminants in the source water, and preemptive and proactive strategies a utility can employ at the treatment plant to deal with both chronic (everyday spillage and leakage) and acute releases (accidental large releases) of AFO contaminants. Cooperation between water suppliers, the agricultural community, regulators and other stakeholders is emphasized. Many CAFO operators want to be "good neighbors", and have no desire to pollute their water supplies. Working in harmony, reasonable communication can be established and sound, cost-effective control measures put in place.

Chapter 1 of this report provides an introduction to the general subject and this research, Chapter 2 discusses the project objectives, and Chapter 3 presents the research methodology. A literature review is provided in Chapter 4 that gives general background information for water utilities to learn about AFOs and CAFOs, the wastes produced at these operations, the fate and transport of AFO-derived wastes in the environment, and the basics of and potential impact of the USEPA CAFO Rule. Chapters 5 through 7 contain the primary focus of this research. Chapter 5 describes potential measures AFOs can take to reduce or contain AFO wastes, and Chapter 6 includes a discussion of strategies water utilities can utilize to reduce the amount of these wastes entering their source water and the impact of these wastes on source water and drinking water quality. Chapter 7 supports the analyses provided in Chapters 5 and 6 by presenting in-depth case studies of several drinking water utilities and how they responded to source water quality concerns presented by AFOs, including a summation of the results of a survey of US water utilities regarding their experiences with CAFOs. The experience of these utilities is particularly valuable for providing example guidance to other water utilities. Chapter 8 discusses the web-based electronic information center that was developed for utilities to readily find detailed information, and to assess their own particular situation. Chapter 9 describes the application potential of this research for the drinking water utility industry, and Chapter 10 lists some potential future research needs.

CHAPTER 2
OBJECTIVES

The primary objectives of this project were to provide information to drinking water utilities regarding AFOs and their waste characteristics, the fate and transport of contaminants from these sources into the environment, potential contaminant impacts on source water quality, and strategies to control and monitor the release of these contaminants in drinking water sources. More specifically, the project was designed to provide guidance to help utilities employ strategies to accomplish the following:

- Identify and characterize AFO waste sources in their watersheds
- Evaluate whether existing animal feeding operations in their watershed are employing proper waste and contaminant reduction and control strategies
- Develop watershed monitoring and control programs
- Develop proactive treatment solutions at the water plant to deal with chronic and accidental AFO contaminant releases
- Become involved in the permitting process to make sure AFOs and CAFOs are required to provide necessary control strategies in the future

Proactive strategies for utilities to address contaminant releases from existing AFOs and CAFOs were identified and evaluated, as well as ways to limit and reduce these contaminants in the future. Strategies for developing partnerships between utilities and agricultural operations are discussed, as are a variety of Best Management Practices (BMPs) for minimizing and managing wastes at AFOs. The research has also identified proactive strategies that the utilities can more directly control, including source water monitoring and protection programs, intake management strategies, incorporating contaminant early warning systems and riverbank filtration or other water treatment processes such as ultraviolet (UV) light disinfection, membrane filtration, and others.

This final report and a web-based information center are the principal deliverables from the project, and are intended to be user-friendly, understandable, and authoritative resources for assisting in the planning and management of drinking water sources. Both are suitable as stand-alone items; however, the objective of the research team was to harmonize these two elements so utilities can most easily find information they need to understand the problems and to develop strategies to deal with AFO impacts on water sources both now and in the future.

The web-based information center and final report can be used together so that users will get the most out of the project findings. One of the functions of the electronic information center is to readily direct the user to alternatives consistent with their utility's needs, capabilities, and goals and that are applicable or effective in the user's situations. Using the website, utilities can investigate these issues without having to wade through a great deal of information that may be applicable to other utilities, but not their particular situation. The web-based information center includes links to electronic files containing extracts from this report, as well as links to other web-based information, including USEPA, US Department of Agriculture (USDA), state agency, and other web sites. A vast amount of information is available from these websites, and the electronic information center is designed to readily lead the user to sources that contain the information they need.

Lastly, this report includes an assessment of the applications potential of this research, as well as missing information and future research needs. Existing information was evaluated to assess whether it is sufficient or conclusive with respect to key issues of importance to water utilities, and to identify areas where future information development is needed.

CHAPTER 3
METHODOLOGY

To address the project objectives described above in Chapter 2, information was obtained from a variety of sources, including the following:

- A comprehensive literature review
- Discussions with utility participants
- Discussions with the project team of expert technical advisors, including a workshop meeting
- Survey of drinking water utilities (including AwwaRF utility subscribers in CAFO-affected areas)
- In-depth examination of utility and CAFO case studies
- On-site visits to water utilities and CAFOs

All data collection efforts conformed to the Quality Assurance Project Plan (QAPP) submitted by the research team and approved by the AwwaRF Project Advisory Committee (PAC) at the beginning of the project (EE&T 2005). Details about some of the key information gathering efforts are outlined below.

PROJECT TEAM

The project team consisted of Environmental Engineering and Technology, Inc. (EE&T) personnel, expert technical advisors, utility participants, the AwwaRF PAC, and participants from other organizations (Table 3.1). A map showing the locations of the project participants is provided as Figure 3.1. The technical advisors were selected based on their expertise in different areas, including dairy cattle, beef cattle, poultry, swine, and aquaculture operations, as well as animal waste treatment. The advisors worked with the project team throughout the project to provide needed information and references, participate in the technical advisor workshop meeting, and perform critical review of all project deliverables.

LITERATURE SEARCH

The following sources of information were used during the project:

- Peer-reviewed literature (journals, books, etc.)
- Federal government literature (USDA, USEPA, USGS, etc.), including the CAFO Rule, other regulations, and documents that contain data, supporting information, and assessment of findings developed by USEPA in support of the CAFO Rule
- State agency literature
- Information available from university, state, and local agricultural extension services
- Information available from professional water (AWWA, AwwaRF, etc.) and agricultural organizations (e.g., cooperative extension offices, the American Society of Agricultural and Biological Engineers, Texas Cattle Feeders Association, etc.)

- Information and critique provided by the project technical advisors and the AwwaRF Project Advisory Committee

Table 3.1
Project participants

EE&T, Inc.
Richard W. Gullick, Ph.D.
Richard A. Brown, P.E.
David A. Cornwell, Ph.D., P.E., DEE
William Cornwell
Ebert Obando

Water utilities
IA – City of Cedar Rapids Water Department
IA - Iowa City Water Division
MN - Mankato Municipal Water Supply
MN - Minneapolis Water Works
MN – Saint Cloud Public Utilities
MN – Saint Paul Regional Water Services
MO - Kansas City Water Services Dept.
NC - Wilmington Water & Sewer
NE - Lincoln Water System
NY – New York City Dept. of Environmental Protection
NY – City of Syracuse Department of Water
OH - City of Columbus Division of Water
OK – Tulsa Metropolitan Utility Authority
PA - Philadelphia Water Department
TX - Amarillo Municipal Water System
UT - Weber Basin Water Conservancy District

Expert technical advisors
Brent Auvermann, Texas A&M University (Amarillo, TX)
Curt Gooch, Cornell University (Ithaca, NY)
Thomas Harter, University of California at Davis (Davis, CA)
Vic Johnson, Kemiron Corp. (Phil Campbell, AL)
Lori Marsh, Virginia Tech (Blacksburg, VA)
Peter Wright, USDA-NRCS (Syracuse, NY)

Other organizations
Delaware County Extension Office, Oklahoma Cooperative Extension Service (Jay, OK)
North Carolina Department of Environment and Natural Resources, Division of Water Quality (Wilmington, NC)
Onondaga County Soil and Water Conservation District (LaFayette, NY)
Source Molecular Corporation (Miami, FL)
Texas Cattle Feeders Association (Amarillo, TX)

AwwaRF Project Advisory Committee
Kim Coy, City of Akron - Water Supply Division (Akron, OH)
Wyndi Klement, Advanced Concepts and Technologies International, LLC (Waco, TX)
Jeff Jollie, USEPA Office of Ground Water and Drinking Water (Washington, DC)

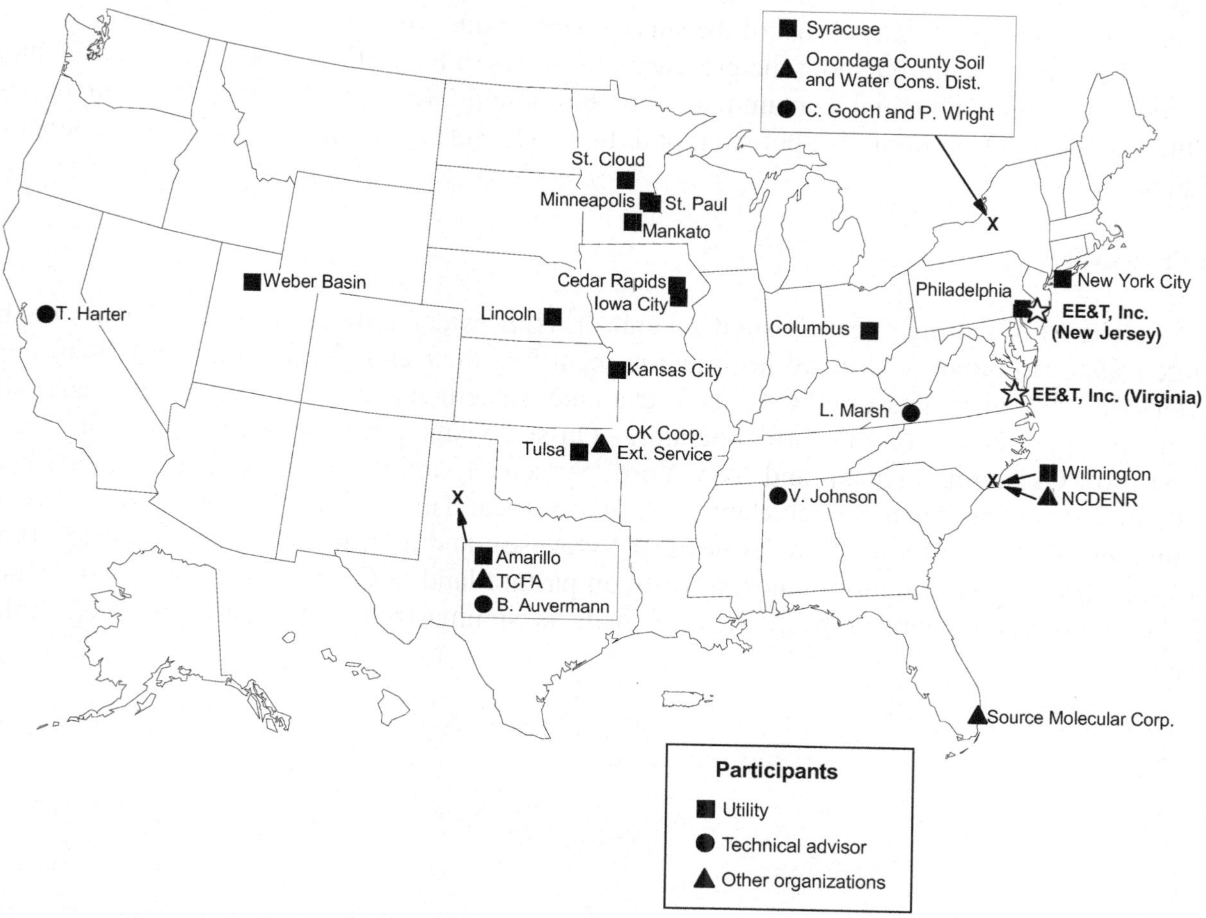

Figure 3.1 Locations of participating utilities, technical advisors, and other organizations

The research team reviewed available published literature, including books and journals, as well as electronic (i.e., internet) database sources. Review by the PAC and input from the technical advisors and utility participants was solicited to provide the best opportunity to make sure the information collected was comprehensive, factually correct, and relevant.

The research team also endeavored to collect and record information from various information sources that may not have been previously documented. This includes data or findings from active studies, which may be documented at some future date, but for which data and preliminary findings have not been adequately disseminated and discussed. An even greater source of undocumented information targeted by this effort was experiences and observations by individuals at water utilities, CAFO farms, and other related organizations that were pertinent to this research. Much of that information is presented in the case study section of this report (Chapter 7).

UTILITY SURVEY

A targeted utility survey was conducted in summer 2005 to gather additional information on drinking water utility experiences with CAFOs and CAFO wastes in different parts of the

United States. The survey was not entirely random, as it was focused on utilities in areas thought more likely to have CAFOs present. However, at least one utility was contacted in each of the 50 states (except for Alaska). Most of the survey participants were AwwaRF subscribers.

The survey inquired about the presence of CAFOs in the utilities' source water area, their concerns regarding potential contamination of the source water, and measures taken by the utilities and CAFOs to minimize potential pollution. The survey results are discussed in detail in Chapter 7.

SITE VISITS

The EE&T project team visited several drinking water utilities in different parts of the United States to obtain additional information regarding their experiences with CAFOs in their watersheds, and visited assorted CAFOs to examine waste management practices first hand and obtain the perspective of the farm operator. On-site visits were made to utilities in Texas (Amarillo), Oklahoma (Tulsa), and New York (Syracuse), and the farms visited included beef cattle in Texas (near Amarillo), chickens in Arkansas (near Tulsa), swine in North Carolina (near Wilmington), dairy cattle in New York (near Syracuse), and a BMP demonstration cattle farm with experimental spreading of chicken waste on pasture land in Oklahoma (near Tulsa). Much of the information obtained from the site visits is summarized in the case study examples presented in Chapter 7.

CHAPTER 4
AFOs AND AFO WASTES

This literature review is intended to provide general background information for water utilities to learn about AFOs and CAFOs, the wastes produced at these operations, the fate and transport of AFO-derived wastes in the environment, the potential impact of these wastes on drinking water source quality, human health and the environment, and the basics of and potential impact of the USEPA CAFO rule. Description of potential measures designed to reduce or contain AFO wastes, as well as strategies water utilities can utilize to reduce the amount of these wastes entering their source water and the impact of these wastes on drinking water quality, are discussed separately in Chapters 5 and 6, respectively.

ANIMAL FEEDING OPERATIONS

Animal feeding operations (AFOs) are located throughout the United States to meet the demand for meat and other animal-related products. The growing trend towards larger operations and concentrating these facilities into smaller areas can magnify waste handling difficulties. Though there are certainly important commonalities, within each industry and farm community there can be a wide range of facilities as well as attitudes, and general waste management practices vary from region to region and farm to farm.

AFOs are agricultural operations where animals are raised in confined situations and feed is brought to the animals rather than the animals grazing in pastures. Specifically, USEPA defines an AFO as an animal feeding operation lot or facility (other than an aquatic animal production facility) where the following conditions are met: (1) animals have been, are, or will be stabled or confined and fed or maintained for a total of 45 days or more in any 12-month period, and (2) crops, vegetation, forage growth, or post-harvest residues are not sustained in the normal growing season over any portion of the lot or facility (for more details see USEPA 2003a).

The largest of these operations that meet definitions outlined in USEPA regulations are called Concentrated Animal Feeding Operations (CAFOs). The number and type of animals and the potential for contaminants from these operations to impact surface water or groundwater determines whether the facility is designated as a CAFO. USEPA estimates there are approximately 257,000 AFOs in the United States (USEPA 2003b). Of those, only 15,500 are considered CAFOs, yet these CAFOs represent nearly 60 percent of the 500 million tons of manure produced annually by operations that confine animals (USEPA 2003a).

Wastes from AFOs can include manure (feces and urine), spent bedding materials (litter), animal parts, dead animals, excess feed, and feed lot runoff, and large operations can produce substantial amounts of waste. The primary pollutants associated with animal wastes are nutrients (primarily nitrogen and phosphorus), pathogens, salts, organic matter, solids, and volatile and odorous compounds (e.g., ammonia and hydrogen sulfide gases, volatile organic compounds (VOCs)). To a lesser extent, animal wastes are also a source of trace elements, antibiotics, hormones, and pesticides. Soaps, surfactants, disinfectants, acids and alkalines, and oil and gas are also potential pollutants from animal operations.

Most of the waste generated at AFOs and CAFOs is ultimately recycled via land application on cropland and pasture. In some cases such as dairy farms and poultry and swine operations, the waste must be removed and either stored long-term or disposed of on a continuous or frequent (e.g., daily) basis. Having waste storage capability allows important flexibility in scheduling land application of manure, litter, and wastewater for times when weather, site conditions, soil and plant nutrient needs, or labor and field availability are more favorable.

Animal manure typically has a high phosphorus-to-nitrogen ratio compared to crop requirements, and using the manure as a fertilizer will relatively over-apply phosphorus compared to nitrogen if the application rate is based on crop nitrogen requirements. In some cases, treatment processes are used to make the waste more amenable for final disposal.

Risks to surface water and groundwater can potentially come from regular activities such as land application of manure wastes and stormwater runoff from animal holding areas, as well as from short-term major events such as the break of a retaining wall for a large liquid manure storage facility and subsequent release of millions of gallons of liquid animal waste into nearby receiving waterbodies. Even standard BMPs such as storing manure and land-applying to dry soil can lead to significant pollution events if a big storm occurs resulting in . The risk of pollution resulting from rainfall on land where manure was spread decreases as the time after spreading increases. Periods of heavy rain can further be problematic in that facilities such as waste lagoons and storage ponds are typically not covered and can overflow.

Information on each state's regulatory programs and initiatives related to AFOs and CAFOs were compiled by Tetra Tech, Inc. (2002). Data on the number of farms for different animal types, as well as estimated production and sales data, can be found through the USDA 2002 Census of Agriculture for each county and state in the US (USDA 2004). An on-line interactive map is available that draws from that USDA database and can be used to identify geographic areas with various amounts of animals or animal farms (www.nass.usda.gov/Census_of_Agriculture/SVG/index.asp). Cornell University also maintains a website where this information can be searched (http://agcensus.mannlib.cornell.edu).

Information on animal production can also be found on the USEPA's National Agriculture Compliance Assistance Center's website at www.epa.gov/agriculture (see www.epa.gov/agriculture/anafopro.html). That website also has links to related information on BMPs, research, regulations, and other pertinent topics. USEPA (2004b) provides a good recent review of the impact of beef cattle, dairy, swine and poultry operations on the environment.

Beef Cattle

In the beef cattle industry, producers typically own most or all of the cattle at their facility. This is unlike integrated industries like poultry and swine where a few large companies own most of the animals. As with the other animal industries, there are a wide range of facility types, so there is no typical operation, just an average one. In general, cattle are held in outdoor pens or corrals and fed and watered until they reach market weight. The sheep and goat industries are similar in production style to beef cattle.

The major cattle producing states are Texas, Kansas, Nebraska, Oklahoma, Colorado, Idaho, Washington, Arizona, and California. Alberta is Canada's top cattle-producing province. Of the approximately 25-million head produced on feed each year in the US, 30 percent are grown within approximately 150 miles of Amarillo, Texas (Auvermann 2005). Increasingly, the industry is becoming dominated by multi-feedyard corporations (for example, owning five to ten feedyards with a total of 350,000 to 600,000 head). One Texas company has over a million head of cattle combined at its different operations.

Figure 4.1 shows the approximate number of beef cattle CAFOs throughout the US with 1,000 or more head (the threshold for classification as a large beef cattle CAFO), and the number of beef farms with over 200 head (medium-size beef cattle CAFOs are classified as 300 to 999 head). Statistics on beef production, collected annually by the National Agricultural Statistical Service (NASS), a branch of the USDA, are available for each state at http://usda.mannlib.cornell.edu/reports/nassr/livestock (USDA 2004). County- and state-level data may also be found in on-line searchable databases at www.nass.usda.gov/Census_ of_Agriculture or http://agcensus.mannlib.cornell.edu.

A fact sheet on CAFO Rule requirements for beef production is available from the Livestock and Poultry Environmental Stewardship (LPES) at www.lpes.org/cafo/12FS_Beef.pdf, and a pictorial description of best management practices for beef cattle production is provided by USEPA at www.epa.gov/agriculture/beef.pdf. Additional information on beef cattle production is available at www.epa.gov/agriculture/ag101/beef.html.

Cattle are generally held in outdoor pens or corrals, with a density of around 150 to 200 square feet per head; thus, approximately 3.5 to 4.5 acres are needed to hold 1,000-beef cattle. As an example, a 45,000-head feedyard covers approximately 200 acres including ancillary areas (feed mill, manure handling, holding ponds, etc.).

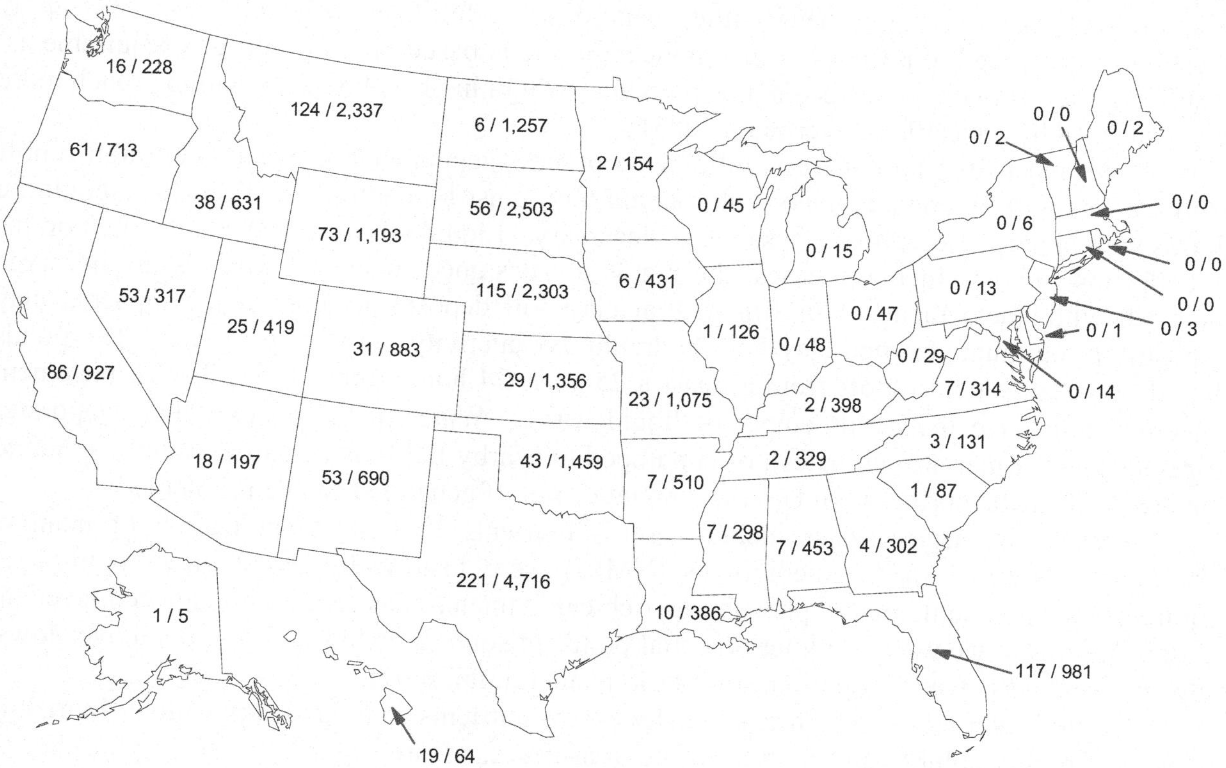

Source: Data from USDA (2004)
Note: Number of large beef CAFOs (>1,000 Head) / Total number of beef AFOs (>200 Head)

Figure 4.1 Estimated number of large beef cattle farms throughout the US

At most sites the cattle are separated in groups of similar age and weight and housed in numerous pens. Feed is given several (e.g., three) times each day to each pen, and records should be kept of the amount of feed applied and consumed to help optimize the feeding regimen. On average, the cattle are fed for 165 days, with a beginning live weight of 350 to 750 pounds. In general, they gain 3 to 4 lbs per head per day eating 20 to 25 lbs of food per day (dry weight), and reach a market weight of 1,100 to 1,300 lbs. Approximately 6 lbs of feed are provided for every pound of weight gain. Each animal may receive up to 3 or 4 different kinds of rations throughout the feeding period, including starter (high roughage), step-up or transitional, and finishing (high energy cereal grain diets; may exceed 90-percent digestibility). Corn, sorghum, distillers' grains, potatoes, and other food sources are used.

The waste management at cattle feedyards can be relatively simple. The predominant waste form is solid manure, though stormwater must also be managed. Drainage ditches uphill of facilities can intercept outside stormwater runoff and channel that flow to downhill of the facility to prevent it from entering the sites and adding to the liquid waste load.

Manure accumulates in each pen during the life of the cattle (approximately four-inches deep). When the cattle are removed for slaughter, the solid manure is scraped up and removed (approximately one-half dry ton per head produced), and is then composted, stockpiled, or hauled off-site for application on nearby agricultural land. The $N:P_2O_5$ ratio of the manure is approximately 1.0 to 1.5, lower than the 2.5 to 3.5 ratio required by cereal crops. Thus, like other animal manure, using beef cattle manure as a fertilizer will relatively over-apply phosphorus compared to nitrogen if the application rate is based on crop nitrogen requirements. Distiller's grain is one type of feed that is particularly high in phosphorus, containing much more than is needed by the cattle (Auvermann 2005).

With most feedyards there is little or no process-generated wastewater. Rainfall runoff from the lots will be contaminated with manure, and must be controlled. Feed lots should be sloped with appropriate drainage systems as necessary. Liquid (stormwater) runoff often drains to collection ponds. In many cases, the ponds are designed to meet USEPA's no-discharge standard, with liners made of *in situ* natural clay soil deposits or imported compacted clay (typically a minimum 18-inch clay with hydraulic conductivity $< 1.0 \times 10^{-7}$ cm/s). The ponds should be sized to hold runoff from at least a 25-year, 24-hour storm event. The management objective is to have the ponds empty of liquid waste. Water in the ponds either evaporates (especially in warmer arid climates), or is pumped to nearby fields as irrigation water. In humid climates such as the northeastern US, the amount of runoff generated can be problematic.

Threats to surface water include excess nutrients from land application of manure. Where used, nutrient management plans (NMPs) were historically based on crop nitrogen requirements, so in some places phosphorus-rich cattle manure has resulted in elevated levels of phosphorus in surface waters. Other potential pollutant sources include holding pond overflows, tailwater from wastewater irrigation, soil erosion, and rainfall runoff.

Groundwater can also be threatened by excess nutrients from land application. Inorganic nitrogen is highly soluble and mobile in the subsurface, as are organic forms of phosphorus. Other threats to groundwater include infiltration of feedyard stormwater into the subsurface, on-site or nearby unprotected wellheads and old well casings, and poor stormwater storage pond liner or construction.

Since the animals are housed outside, beef cattle CAFOs are more susceptible to weather conditions than some other operations such as swine and poultry. Large storms can result in a substantial amount of runoff contaminated with animal wastes, and thus proper design of stormwater drainage systems and settling basins is especially important. The applicable regulatory requirements do not always conform well to the weather conditions that are encountered, and thus very large storm events can potentially be problematic.

Some current trends in the US beef cattle industry include the following (Auvermann 2005):

- Steady growth and expansion, but not many new feedyards
- Above-average growth in semi-arid West
- Accelerated phosphorus-based nutrient management planning
- Net flow of phosphorus from mines in Florida (used as fertilizer base) to the corn belt (where it is used as crop fertilizer) to cattle-feeding states (where it is disposed of in manure on croplands and pastures)
- Over time, declining water tables in the High Plains will reduce irrigated acreage, crop yield and sustainable fertilization rates
- Animal mortality disposal a growing concern
  - Rendering is a vanishing prospect for many remote producers
  - Burial may not be allowed much longer in many places
  - Incineration is energy intensive
  - Biosecurity issues with dead carcasses
- Fugitive dust showing up as a source of soil nutrient enrichment downwind of cattle feedyards
- Airborne ammonia dissolves readily in downwind surface water; airborne emissions may eventually be regulated due to secondary particulate matter production
- Energy production from manure and manure products is increasing

Dairy Cattle

Dairy farms are comprised of housing facilities for lactating cows, dry cows, and replacements; cropland to grow forage-based crops like corn silage and haylage; storage facilities for these forages (silages) and also purchased feeds; and manure treatment and storage facilities. Some farms have separate facilities for milking cows, called milking centers. Cows are typically milked two or three times per day, and manure collection and removal is generally a steady process. Manure is rich with organic nitrogen and phosphorous and thus is recycled to cropland as fertilizer.

Sources of pollution from dairy operations can include manure storage and recycling of manure on cropland, as well as milking center wastewater, barnyard runoff, and silage leachate. The amounts and concentrations of these sources vary daily, seasonally, and from farm to farm. Silage leachate typically has higher concentrations of nitrogen, phosphorus, and biochemical oxygen demand (BOD) than milk house washwater and barnyard runoff, but less than manure. The nitrogen, phosphorus, and BOD concentrations in silage leachate vary depending on timeliness of crop harvest, silo design and management, and presence of dilution water.

The major dairy producing states are California, Wisconsin, New York, Pennsylvania, and Idaho (USDA 2004). Figure 4.2 shows the approximate number of medium (200 to 699 head) and large (≥700 head) dairy cattle CAFOs throughout the US. A pictorial description of best management practices for dairy cattle production is provided by USEPA at www.epa.gov/agriculture/dairy.pdf, and an LPES fact sheet on CAFO Rule requirements for dairy barnyards is available at www.lpes.org/cafo/11FS_DairyBrn.pdf.

A list of links to websites with various resources for dairy AFO operations is available at http://groundwater.ucdavis.edu/dairy_links.htm. Cornell University maintains a website with information on manure management (www.manuremanagement.cornell.edu), and a good report on managing dairy manure is available from the University of California (2005). Cost-effective dairy manure management practices are discussed further by the National Dairy Environmental Stewardship Council (NDESC 2005). Statistics on dairy production are collected by the USDA NASS for each state (USDA 2004). County- and state-level data may also be found at www.nass.usda.gov/Census_of_Agriculture or http://agcensus.mannlib.cornell.edu.

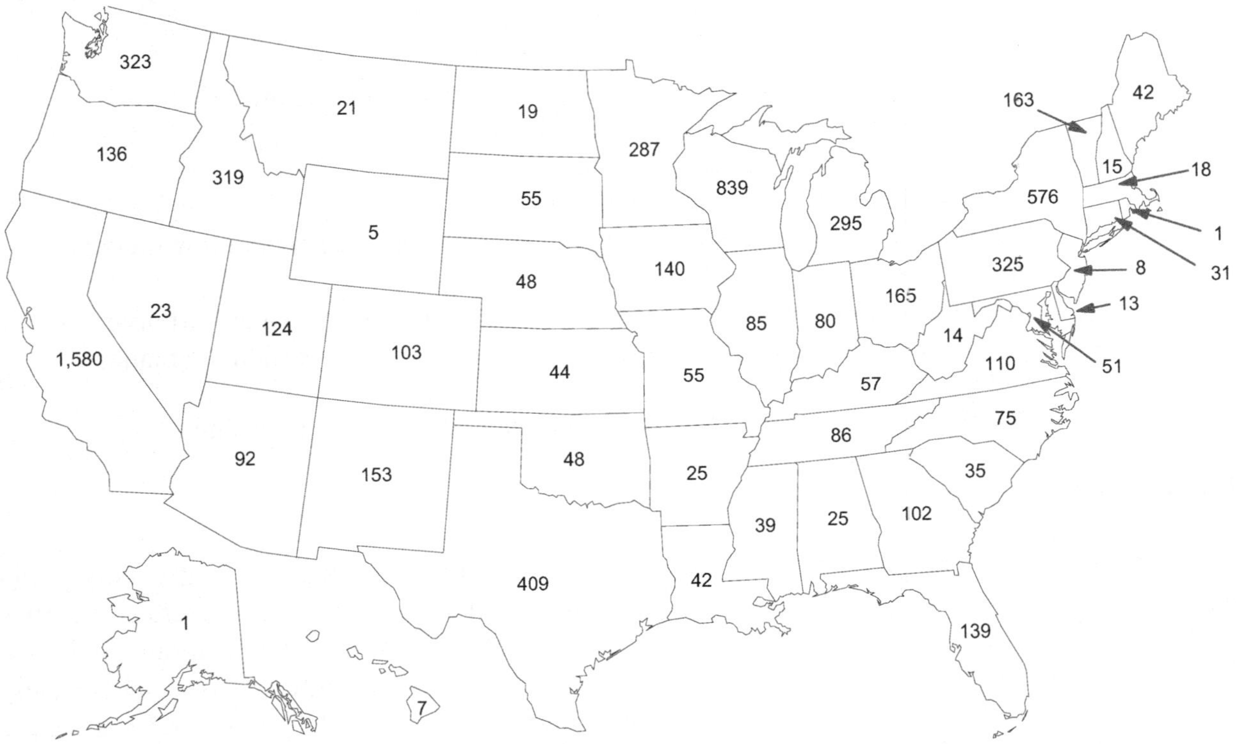

Source: Data from USDA (2004)

Note: Combined total number of medium (200-699) and large (700 or more head) dairy cattle CAFOs

Figure 4.2 Estimated number of medium and large dairy cattle CAFOs throughout the US

A comprehensive nutrient management plan (CNMP) will include not only manure land application schedules, but means to control other concentrated sources and smaller sources as well. Strategies include elimination, reduction, and reuse of waste sources. Concentrated waste flows should be collected, and relatively-dilute high flows can be treated with a grass filter in most states. General primary solutions include runoff control in production areas, and spreading manure on land in accordance with a CNMP. Manure storage facilities help allow for flexibility in spreading schedules to accommodate weather and climatic conditions, and roofed structures help prevent contamination of runoff. Storage facilities and treatment lagoons are typically designed to store (and treat for lagoons) silage juice, manure, and direct rainfall and runoff rainfall, as well as also be able to contain at any time rain from a 25-year, 24-hour storm event. Another important measure is the collection and isolation of calf wastes separate from that of other animals, since calves are a primary source of *Cryptosporidium* oocysts and *Giardia* cysts. Maintaining calf health is also important for minimizing the presence of the pathogen.

Substantial quantities of fresh manure can be deposited on earthen barnyards. One major source water protection measure that can be done with some dairy operations is to eliminate the barnyard. Other potential measures include reducing the size of barnyards, reducing the amount of runoff, controlling runoff flow, and using sorptive or vegetative buffers to help treat runoff as it leaves the barnyard. If possible, open barnyards may be replaced with a concrete pad to better collect waste, and a roof to keep precipitation from contacting the manure. This will reduce not only runoff of manure constituents, but also soil erosion. Depending on the purpose of the barnyard (holding, feeding or watering, resting, exercise, heat detection), farmers may be willing to replace it with a concrete pad or shelter, especially if financial assistance for installing these BMPs are available. Barnyards should be sized correctly, and concrete pads can simplify manure collection and prevent infiltration to groundwater. Roofs for barnyards prevent precipitation from directly falling on the manure.

Vegetative buffer strips such as grass filters can provide low-cost efficient treatment of both barnyard runoff water and milking center wastewater, but may require some maintenance. Waste treatment can include aerobic and anaerobic biodegradation. Aerobic is faster, producing fewer odors than anaerobic biodegradation, but can be energy intensive. Anaerobic digestion is an increasingly popular option for treating cow manure.

Poultry

Poultry production in the United States falls into three primary categories:

- Broilers, which are raised for meat
- Layers, which produce both table eggs and eggs for replenishing both the layer and broiler flocks
- Turkeys, which can be further subdivided into meat birds and egg layers that replenish turkey flocks

Statistics on poultry production collected by the NASS (USDA 2004) are available for each state at http://usda.mannlib.cornell.edu/reports/nassr/poultry/pbh-bbp/. Statistics for county- and state-level data may also be found in searchable databases at www.nass.usda.gov/Census_of_Agriculture or http://agcensus.mannlib.cornell.edu. A pictorial description of best management practices for poultry production is provided by USEPA at www.epa.gov/agriculture/poultry.pdf, and an LPES fact sheet on CAFO Rule requirements for poultry production areas is available at www.lpes.org/cafo/13aFS_Poultry.pdf.

Figure 4.3 shows the approximate number of layer hen farms (with >20,000 hens at least 20-weeks old) throughout the US. Figure 4.4 shows the approximate number of poultry broiler farms throughout the US (a minimum number of birds for these farms was not designated by the data source). Medium-size layer hen or broiler CAFOs using liquid manure handling systems are classified as having 9,000 to 29,999 chickens, and large CAFOs have 30,000 chickens or more (Table 4.2). Medium-size layer hen CAFOs using non-liquid manure handling systems (e.g., dry-litter operations) are classified as having 25,000 to 81,999 layers, and large CAFOs have 82,000 layers or more. Medium-size chicken broiler CAFOs using non-liquid manure handling systems (e.g., dry-litter operations) are classified as having 37,500 to 124,999 birds, and large CAFOs have 125,000 broilers or more.

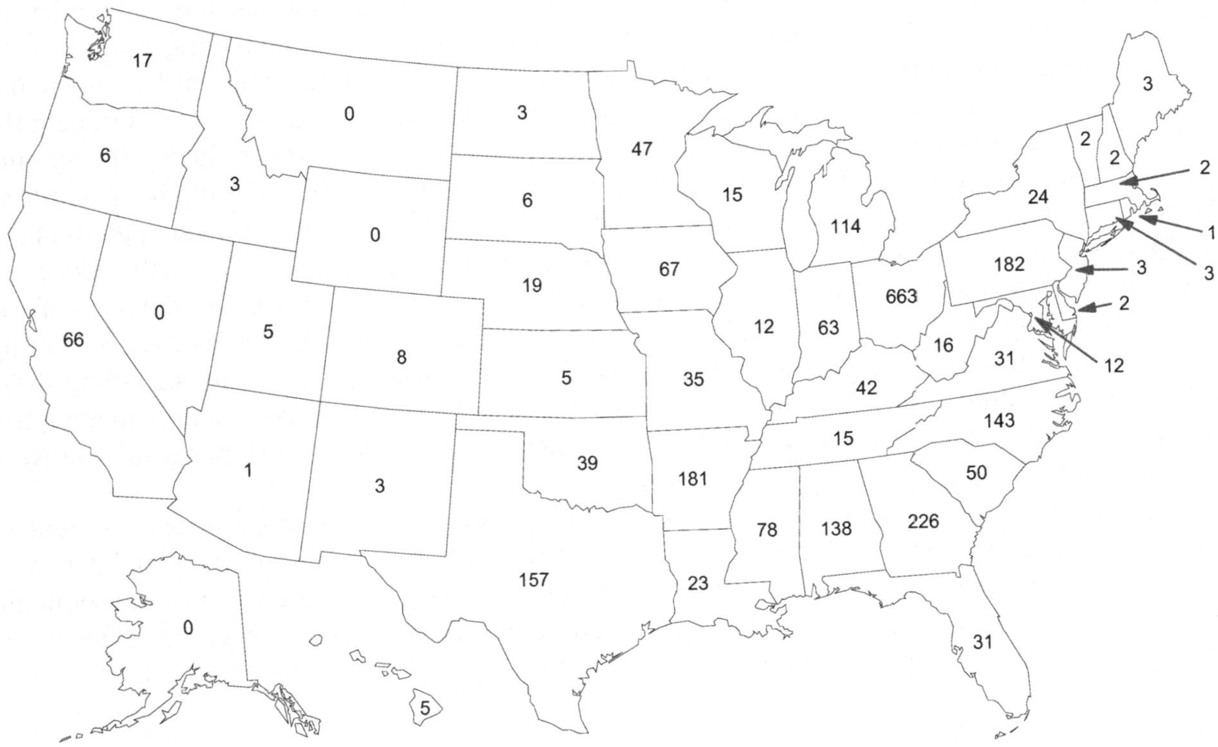

Source: Data from USDA (2004)
Note: Total number of farms with >20,000 layer hens, 20 weeks or older

Figure 4.3 Estimated number of layer hen farms throughout the US

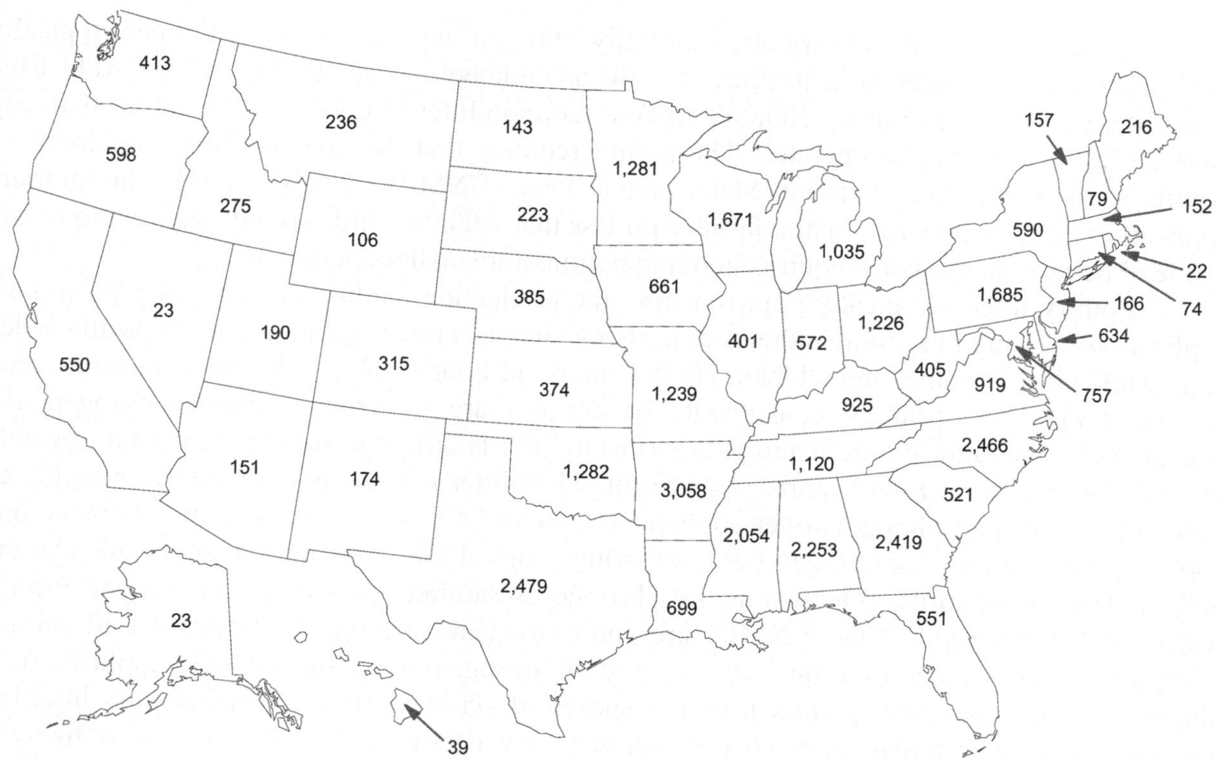

Source: Data from USDA (2004)
Note: Total number of broiler and other poultry meat-type farms

Figure 4.4 Estimated number of broiler and other poultry meat farms throughout the US

Broiler Production

Broiler production is an integrated industry. This means that the birds are owned by the company that ultimately processes and markets them. Examples of poultry integrators include Tyson, Perdue, and Pilgrim's Pride. While some farms are owned by the integrator, many are privately owned and the integrator pays the farmer to raise the birds. Broiler production facilities tend to concentrate near poultry processing facilities. The six largest broiler producing states are Arkansas, Texas, North Carolina, Georgia, Alabama, and Mississippi. Many other states also have some large producing farms, and, as mentioned previously, production tends to be concentrated in areas near processing facilities. The Shenandoah Valley in Virginia and the Delaware-Maryland-Virginia peninsula are examples of concentrated production regions.

Virtually all broilers are raised in confinement, i.e. within an environmentally-controlled structure. Birds are typically raised on the floor of the structure, where they are allowed to move freely. The floor is covered with an absorbent, high-carbon material, which is referred to as litter. Example litter materials include sawdust, wood shavings and peanut hulls. The specific material used depends upon what is available locally. Litter serves as a bedding material and to absorb moisture. It also serves as manure storage. The clean-out schedule for the used litter varies by integrator, and can range from as frequently as after each flock (approximately eight weeks) to several flocks (six months to a year).

Used litter is rich in nutrients, especially nitrogen and phosphorus. Farmers typically land apply the used litter as a fertilizer or soil amendment. The 2003 USEPA CAFO Rule designates poultry farms raising 100,000 birds or more on litter as CAFOs and requires that they have an NPDES permit to operate. The permit requires that the operator have and follow a certified Comprehensive Nutrient Management Plan (CNMP), which controls the nutrient application rates to the land. While farms with less than 100,000 birds are not required to obtain a federal permit, many states require operating permits for smaller operations.

Poultry litter is typically removed from the production facility and stacked prior to land application. During this storage, the litter heats up, and as a result pathogens are typically killed (e.g., McKinley, Broome, and Oldham 2000; Carter and Poore 1995). Therefore, once the litter is land applied the primary constituents of concern are nutrients, primarily nitrogen and phosphorus. The ratio of nitrogen to phosphorus in litter is different from the ratio of nitrogen to phosphorus that most crops require. As a result, when litter is land applied to meet the nitrogen needs of a crop, phosphorus is over-applied by a factor of four to five. Because farmers land applied poultry litter on a nitrogen basis for many years, those fields tend to have soils that are very high in phosphorus. Once soils are phosphorus saturated, losses to surface and ground water are accelerated. New CNMPs are more frequently being developed based on the phosphorus requirement of crops (as opposed to nitrogen) to avoid an over-application of phosphorus. In some cases poultry litter is disposed of via incineration. Since poultry litter has lower moisture content than most other animal manure wastes (see Table 4.5), it is more likely to be incinerated compared to the other manures.

Egg Production

The egg production industry is somewhat different from broiler production. While it is not an integrated industry, approximately 95 percent of the table eggs produced in the United States come from only 200 large facilities. The top six egg-producing states in 2004 were Iowa, Ohio, Pennsylvania, Indiana, California, and Georgia (http://usda.mannlib.cornell.edu/reports/nassr/poultry/pbh-bbp/). Data for these and other states can be found at the same website.

Housing and waste handling systems for layers differ from that used for broilers. Layers are typically confined to cages to make egg collection easier. Their manure falls below the cages, where it can be captured on a belt and removed from the building daily, allowed to fall to a pit below the building where it accumulates and is cleaned out periodically, or "flushed" by running water through troughs under the cages. Flushing systems typically have lagoons for receiving and treating the waste. The marked difference between these systems and broiler production is the lack of a carbonaceous bedding material. Like broiler production, the waste from layer systems is ultimately land-applied, and the constituents of concern are primarily phosphorus and nitrogen.

USEPA recognizes that there is greater potential for environmental harm from laying hen manure than from broiler litter. This is evident in the federal definition of CAFOs for layers. While it takes 125,000 broilers on dry litter to be designated a CAFO, 30,000 birds on a flush system or 82,000 layers on a non-liquid manure handling system are designated CAFOs. Again, CAFOs are required to have a NPDES permit, which includes a comprehensive nutrient management plan. The vast majority of egg production is large enough to fall under the CAFO Rules.

Turkey Production

Turkey production is primarily for meat, but there are operations that produce eggs to replenish the meat stock. Like chickens, the meat birds are raised on litter and the egg-producers are caged, with manure handling systems similar to laying chickens. The turkey industry is integrated just like the broiler industry, with large companies that own the processing facilities also owning the birds. Again, private farms often grow the birds under a contract with the integrator. USEPA defines a turkey CAFO as an operation with 55,000 or more birds. Again, the litter or manure is ultimately land applied and the constituents of concern are primarily phosphorus and nitrogen. The top five turkey producing states are Minnesota, North Carolina, Arkansas, Missouri, and Virginia (http://usda.mannlib.cornell.edu/reports/nassr/poultry/pbh-bbp/). Data for these and other states can be found at the same website.

Swine

The United States swine industry operates predominately under a "multi-site" model, meaning that different ages and stages of production tend to occur on different farms, including the following:

- Breeding and gestation (where sows are impregnated, typically by artificial insemination, and kept until they are ready to give birth)
- Farrowing (sows ready to give birth are moved to a farrowing unit, where they remain until the piglets are ready to wean)
- Nursery (weanlings to 55 pounds)
- Grow-out (55 pounds to market weight, about 225 pounds)

The vast majority of swine are produced under an integrated model, meaning that the pigs are owned by a company that also owns the processing facility. Often the integrator company will contract with private farms to rear the pigs; hence the private farmer is following protocol (facility design, feeding, etc.) specified by the integrator, and integrators also generally supply the feed.

The USDA NASS reports the US hog inventory (by state) four times per year, and the most current statistics can be found at http://usda.mannlib.cornell.edu/reports/nassr/livestock/php-bb/. The top seven swine producing states are Iowa, North Carolina, Minnesota, Illinois, Indiana, Missouri and Nebraska. Data for these and other states can be found at the same website. Figure 4.5 shows the approximate number of swine farms with >500 head throughout the US.

The USEPA defines a swine operation with 2,500 or more swine over 55 pounds, or 10,000 or more swine under 55 pounds, as a large CAFO. Operations with 750 or more swine over 55 pounds or 5,000 or more under 55 pounds and that use a man-made ditch or pipe to carry manure or wastewater from the operation are designated as medium CAFOs. All CAFOs must have a permit from their permitting authority (typically the state environmental protection agency), which designates how manure is handled. Swine production statistics for each state are available from the NASS at http://usda.mannlib.cornell.edu/reports/nassr/livestock/php-bb/. Statistics for county- and state-level data may also be found in searchable databases at www.nass.usda.gov/Census_of_Agriculture or http://agcensus.mannlib.cornell.edu.

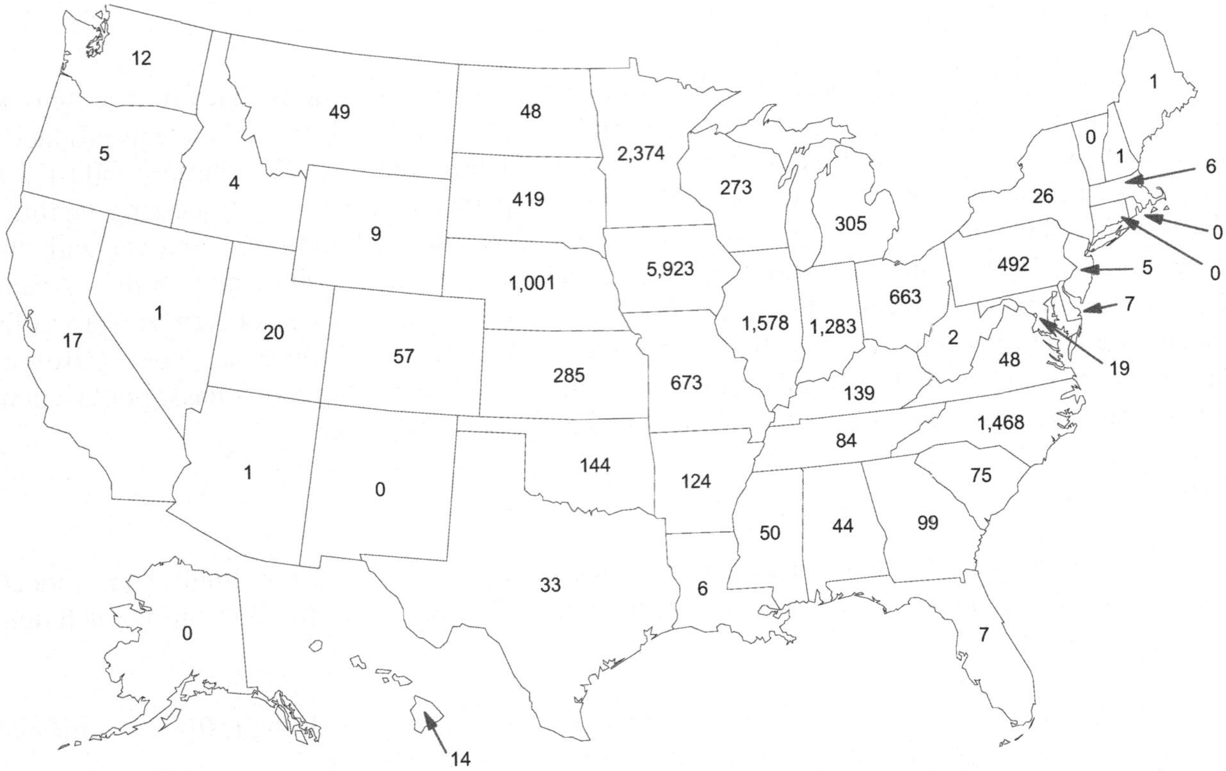

Source: Data from USDA (2004)
Note: Swine farms with >500 head

Figure 4.5 Estimated number of swine farms with >500 head throughout the US

Swine CAFO permitting requirements are summarized by USEPA at www.epa.gov/npdes/pubs/cafo_brochure_swine.pdf. An LPES fact sheet on CAFO Rule requirements for large swine operations is available at www.lpes.org/cafo/13bFS_Swine.pdf. A pictorial description of best management practices for swine production is provided by USEPA at www.epa.gov/agriculture/swine.pdf.

The housing and production system, including the waste handling system, depends on the swine growth stage, and to some extent on geographic location. Most swine are raised in total confinement, over slatted floors. Manure falls through the floor slat into a pit located below. In warmer climates, the pit is typically shallow, and the manure is flushed out of the building into a lagoon. Water is drawn off the lagoon to flush the system; i.e., it is a closed loop. Lagoons are engineered to act as anaerobic treatment units and include volume for long-term storage of sludge (typically 5 to 15 year's worth of sludge-storage volume). The waste lagoons can also be a source of ammonia to both air and water.

Treated lagoon effluent (lagoon supernatant) is periodically land applied to meet nutrient requirements of crops. The constituents of concern are primarily nutrients, specifically nitrogen and phosphorus. Perhaps the greatest waste-handling challenge to a facility with a lagoon occurs when the sludge-storage volume is filled and sludge must be removed. This material is significantly more nutrient dense than the liquid effluent and requires a large land area on which to apply it.

Periods of heavy rain can also be problematic, as the lagoons are typically not covered and can overflow. Furthermore, spraying of lagoon water on saturated fields is not recommended and leads to much more waste runoff into surface waters. Thus, during very rainy periods, swine farm operators are sometimes left with a choice of spraying on wet fields or taking the risk of allowing the lagoons to overflow (generally, overflows should always be avoided).

In cooler climates, lagoons tend not to function well during extended cold periods and therefore are not commonly used in these locations. Instead, the pit under the slatted production floor tends to be much deeper and serves as manure storage. The pit is pumped periodically, and the manure is land applied. Another production strategy in cold climates involves non-slatted floors. In this system, the manure is frequently scraped into a holding pit, from which it is land applied as needed for crop fertilization or when weather conditions permit (non-saturated and non-frozen ground).

Yet another production system, which is used for grow-out, is called deep-pack or hoop house production. This system is less common than slatted floor or scrape systems, but is gaining in popularity in some areas due to its low initial cost for facility construction. Under this production model, swine are reared in a greenhouse-like structure (typically a Quonset-type structure with a vinyl or other flexible covering) and a bedding material is placed on the floor. Somewhat similar to poultry broiler production, the bedding material serves to absorb liquids and as manure storage. Bedding is added periodically during production, and removed at the end of the production cycle. Bedding material is also land applied.

Since the ultimate fate of swine manure is usually land application, there is cause for concern that nutrients and pathogens can be transported to ground and surface waters during storm events. Die-off rates for pathogens tend to be high after land application, but a storm event immediately following spreading could result in some pathogen loading to near-by water bodies. Best management practices to protect water quality include proper feeding strategies to minimize nutrient concentrations in manure, careful operation of the waste handling system to prevent spills and overflows, and proper land application according to a comprehensive nutrient management plan (required of CAFOs and for farms seeking cost share programs from the USDA Natural Resource Conservation Service).

As with all animal farming operations, dead animals must be properly disposed of. Dead hogs are either buried, incinerated, or 'rendered' (i.e., taken away for disposal by a third party which grinds them up for use in animal feed, as a base material for cosmetic products, and other uses). Dead hogs should be removed from the pens as soon as they are found, and are often temporarily stored in a nearby dumpster, or 'dead box' (also known as an Animal Disposal Unit (ADU) container), which may be checked as often as once each day by whomever removes them from the farms. Farms should have contingency plans prepared for disposal of large quantities of animals in the event of a major incident causing large-scale mortality.

The swine industry, like poultry, dairy, and beef, is challenged by the fact that the phosphorus content of the manure relative to the nitrogen content is much higher than the phosphorus-to-nitrogen ratio needed by most crops. Because swine farmers land applied manure based on the nitrogen needs of crops for many years, there has been a tendency for phosphorus to build up in soils. This can result in phosphorus losses to surface and ground waters. For some watershed and states, new regulations require land application rates limited by the potential for phosphorus losses to surface and ground waters.

Aquaculture

The United States Department of Commerce defines aquaculture as the propagation and rearing of aquatic organisms in controlled or selected aquatic environments for any commercial, recreational, or public purpose. Due to declining natural fish and shell fish populations and an increase in the demand for fish and seafood, aquaculture has become the fastest growing sector of American agriculture. Aquacultural production in the US includes the production of food fish, ornamental fish, baitfish, mollusks, crustaceans, aquatic plants and algae, and some reptiles such as alligators and turtles. Production occurs in both fresh and salt water and employs a number of different production systems.

According to the 1998 USDA Census of Aquaculture, there were slightly over 4,000 aquaculture farms in the United States (USDA 1998). Like other animal production industries, the trend appears to be toward highly concentrated production systems. The aquaculture census revealed that of the 4,000 plus farms reporting, 208 reported sales of over $1 million each. This shows that 5 percent of the farms were responsible for over 60 percent of the total US aquaculture sales in 1998. Figure 4.6 shows the approximate number of food fish farms throughout the US in 1998.

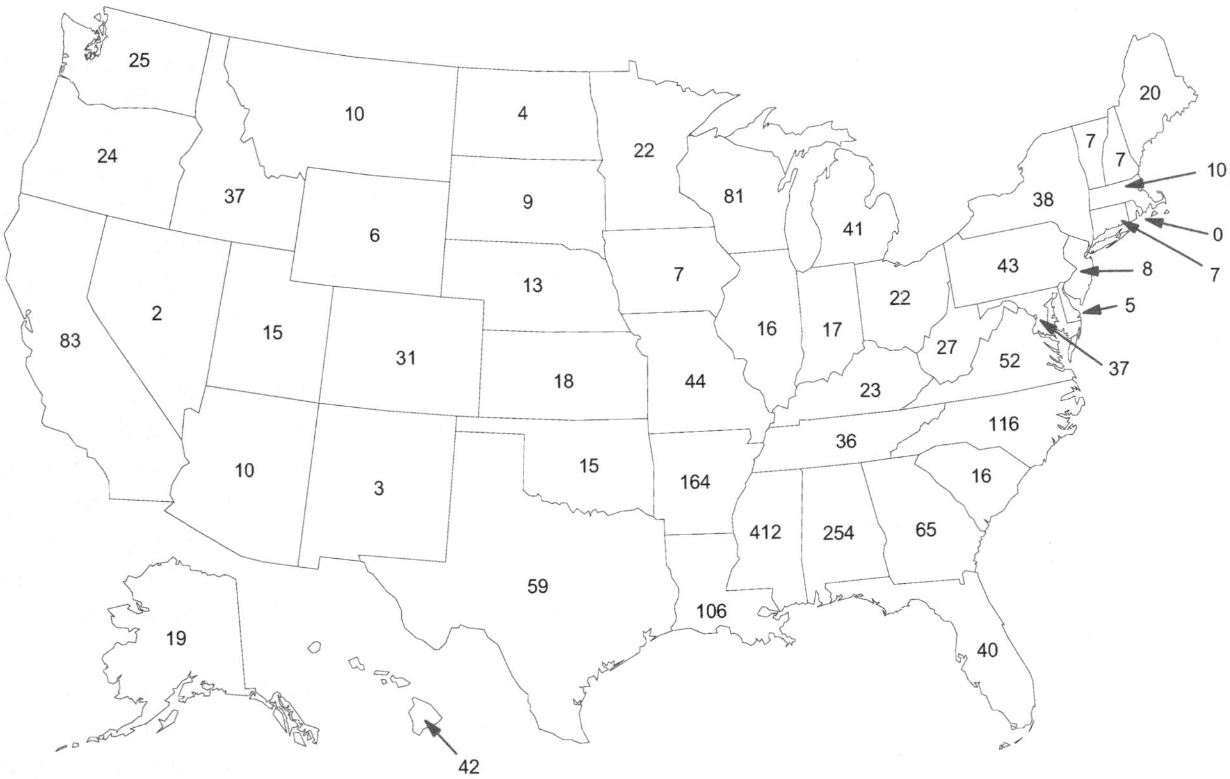

Source: Data from USDA (1998)
Note: Does not include bait fish or ornamental fish

Figure 4.6 Estimated number of food fish farms throughout the US

The most recent aquacultural production statistics can be found in the 2002 Census of Agriculture, which presents aquacultural products sold by category (catfish, trout, bait fish, etc.) and by state within each category (USDA 2004). Statistics for county- and state-level data may also be found in searchable databases at www.nass.usda.gov/Census_of_Agriculture or http://agcensus.mannlib.cornell.edu. The following information was extracted from that database:

- The catfish industry is by far the largest sector of American aquaculture, accounting for almost half of all aquacultural sales. Catfish are produced almost exclusively in ponds. Mississippi accounts for almost one-third of the US production. Alabama has the next highest production, followed by Arkansas and then Louisiana. These four states accounted for over 96 percent of the US catfish production.
- The second largest aquaculture industry in the US is trout. Trout are most commonly produced in raceways, or flow-through systems. Idaho is the major trout producer, accounting for almost 35 percent of all production in 2002. The next largest producing states were Arkansas and North Carolina, accounting for about 10 percent of the national production each, then California and Pennsylvania, accounting for just under 3 percent of the national production each.
- Following catfish and trout, in terms of pounds produced annually in the United States, is "other food fish", which include hybrid stripped bass, tilapia, and yellow perch, among others. Based upon farms reporting in pounds produced, the top six producing states in order of production are Washington, California, Texas, Virginia, Mississippi and North Carolina.
- Baitfish represents the next largest aquacultural product in the United States. Of the farms reporting on a pounds-sold basis, Arkansas produced almost 65 percent of all baitfish sold in the US, with Minnesota producing approximately 13 percent.

The primary water quality variables of concern in aquaculture effluent are nutrients (nitrogen and phosphorus), suspended solids, and biochemical oxygen demand. Catfish pond effluents generally have higher concentrations of nutrients and organic matter than natural stream waters but much lower concentrations than municipal and industrial wastewater (SRAC 1999). For pond systems, which represent the greatest percentage of aquacultural production systems in the US, discharges typically occur when ponds are drained during harvest, during heavy rainfall events, and occasionally when water is discharged to maintain good water quality (by introducing clean makeup water).

Aquaculture facilities are not subject to the CAFO Rule, as USEPA regulates them in a separate point source category (Animal Production Point Source Category), requiring an NPDES permit if they meet the following criteria (40 CFR Part 122 Appendix C):

1. Contains, grows, or holds fish in raceways, ponds, and other similar structures, and
2. Discharges pollutants to surface waters of the United States during at least 30 days per year

AND

3. Meets any of these criteria for production or feeding:

i. Is a cold water aquaculture facility that has a production of 20,000 pounds or more of cold water fish per year; and feeds 5,000 pounds or more of food in the month of maximum feeding

OR

ii. Is a warm water aquaculture facility that has a production of more than 100,000 pounds of warm water fish per year

In addition to the above, any facility may be required to have an NPDES permit if USEPA or the primacy state determines it to be a significant contributor of pollution to waters of the United States. Facilities that are not required to hold NPDES permits must still meet applicable state requirements, which vary by state. Some states regulate discharges from aquacultural facilities as industrial waste, while others treat it as agricultural waste. Typically agricultural waste standards are not as stringent as industrial waste standards.

Effluent limitations guidelines (ELGs) were promulgated for the Concentrated Aquatic Animal Production Point Source Category in August 2004 (40 CFR 451, USEPA 2004c). These ELGs, like the ELGs for CAFOs, were promulgated to reduce discharges of pollutants, specifically targeting releases of total suspended solids, nutrients, and chemicals, or pharmaceuticals used at these facilities.

Aquaculture facilities that meet the above definitions are required to prevent or control accidental releases of drugs, pesticides, or other chemicals. In particular, these facilities must monitor and control feeding of the animals to limit discharges of excess feed, including requirements for active monitoring and development of feed management strategies for open water systems. Open water systems are further required to develop and implement strategies to prevent releases of dead aquatic animals into wastewater. The wastewater treatment systems at these facilities must be maintained and staffed by trained personnel. Facilities must retain records of number and weight of animals and amount of feed used, as well as operation and maintenance records of the facility wastewater treatment systems. Overall, the facility must develop and maintain a Best Management Practice plan describing how the facility will meet all of these requirements. A USEPA fact sheet on the requirements for aquaculture is summarized at www.epa.gov/waterscience/guide/aquaculture/fs-final.htm.

USEPA describes aquaculture operations and related BMPs at www.epa.gov/agriculture/anaqubmp.html. The Aquaculture Network Information Center (AquaNIC; http://aquanic.org/) provides a large number of web links to various information sources related to aquaculture resources. The Southern Regional Aquaculture Center (SRAC, www.msstate.edu/dept/srac/) provides a series of fact sheets about aquaculture and aquaculture waste management. Aquaculture systems are described by the Idaho Department of Environmental Quality at www.deq.idaho.gov/water/prog_issues/agriculture/aquaculture.cfm. Waste management systems for aquaculture facilities, including BMPs to reduce aquaculture waste, are summarized by Miller and Semmens (2002), and discussed in more detail by Idaho DEQ (2002).

As an example, the Southern Regional Aquaculture Center offers the following suggestions for managing pond effluents:

- Use high quality feeds and good feeding practices to minimize waste production
- Provide adequate aeration to promote maximum feed conversion
- Minimize water exchange
- Operate ponds for several years without draining
- Manage the pond water level to capture rainfall
- Allow solids to settle before discharging

- Optimize watershed areas so that ponds do not produce runoff during rainfall events
- Treat effluent in constructed wetlands
- Use effluents to irrigate terrestrial crops

Following ponds, raceways are the next most common production system. Raceways typically divert water from a stream, allow the water to cascade from one production tank to the next and then return the water to the stream. Because dissolved oxygen is typically the limiting factor for fish production, tanks are often aerated by cascading water or adding oxygen. Depending on the size of the operation and the specific state regulations, various waste treatment strategies may be employed in the system. As with a pond system, waste management begins with feed management, since feed is the source of most of the waste products. Larger systems typically involve some method of solids removal, such as settling to a central drain, or using a settling basin. Filtration is also possible using drum, disk, bead or sand filters. Once removed, solids can be composted or land applied.

USEPA CAFO RULE

The USEPA published their Final Rule (USEPA 2003a) regarding Animal Feeding Operations (AFOs) and Concentrated Animal Feeding Operations (CAFOs) on February 12, 2003, and it became effective as of April 14, 2003. The purpose of the CAFO Rule was to establish regulations that will benefit human health and the environment. The rule revises and clarifies the section of the Clean Water Act dealing with CAFOs and is intended to ensure that CAFOs take proper action to protect the nation's water quality (40 CFR Parts 9, 122, 123, and 412; National Pollutant Discharge Elimination System Permit Regulation and Effluent Limitation Guidelines and Standards for Concentrated Animal Feeding Operations (CAFOs); Final Rule).

The CAFO Rule is intended to simplify and clarify the existing regulations, and requires CAFOs to: (a) be permitted as a point source; (b) develop plans for storing, using, and disposing of animal wastes; and (c) to prepare annual compliance reports on waste disposal practices. It maintains the current NPDES permitting structure (Large, Medium, or Small-size operations) and the criteria for when USEPA or the state define or designate a small or medium AFO as a CAFO. The rule established a requirement for all CAFOs to apply for an NPDES permit, and to develop and implement a site-specific nutrient management plan that include appropriate best management practices to protect water quality. Facilities may elect to use innovative technologies and alternative management practices. The regulated operations include several types that were previously not considered CAFOs, including large facilities that discharge only as the result of a large storm event, stand-alone immature swine or heifer operations, and large "dry-litter" poultry operations (this latter requirement was subsequently successfully challenged in court; see discussion below of the 2005 Waterkeeper Alliance Inc. et al. v. USEPA case).

The rule includes performance expectations for existing and new sources to ensure proper storage of manure and wastewater at CAFOs, as well as for proper land application practices. These requirements are intended to ensure these sources meet the standards set up by the Clean Water Act's effluent limitation guidelines and standards (ELGs). The rule maintains or strengthens the previous requirements for reducing pollution discharges from production areas and promotes innovation. Facilities must provide storage that will contain their manure plus the precipitation from a major storm. The types of operations affected by the rule are listed in Table 4.1.

The NPDES permitting structure that separates small, medium, and large CAFOs is used to define operations under the CAFO Rule. The explicit definitions of large, medium, and small CAFOs are shown in Table 4.2 (from USEPA 2003a). Large facilities are considered CAFOs if they fall within the size range provided in §123.23(b)(4) of the USEPA CAFO Rule (USEPA 2003a). Medium AFOs are defined as CAFOs only if they fall within the size range provided in §122.23(b)(6) and they meet one of the two specific criteria governing the method of discharge: (1) Pollutants are discharged into waters of the United States through a man-made ditch, flushing system, or other similar manmade device; or (2) pollutants are discharged directly into waters of the United States that originate outside the facility and pass over, across, or through the facility or otherwise come into direct contact with the confined animals. Small facilities are CAFOs only if they are so designated by USEPA or the state NPDES permitting authority.

The focus of the CAFO Rule is on the largest operations and the operations that pose the greatest environmental risk. However, the USEPA also designed the CAFO rule to support and complement the various voluntary and other programs implemented by the USDA, USEPA, state agencies, and other organizations that help the vast majority of smaller animal feeding operations not addressed by this rule (USEPA 2003a).

For certain operations in some states, similar regulatory programs already existed and thus the CAFO Rule did not have a substantial impact. As examples, poultry operations were already regulated at the state level in Virginia, so the CAFO Rule isn't having much impact there. Texas already regulated their beef industry, thus most of the impact of the CAFO Rule has been in terms of clerical and record-keeping requirements. In contrast, the beef industry in Colorado was not nearly as regulated as that in Texas, and there has been a lot of change in Colorado in response to the CAFO Rule.

Table 4.1
Examples of entities regulated by the USEPA CAFO Rule (USEPA 2003a)

Examples of regulated entities	North American industry code (NAIC)	Standard industrial classification code (SIC)
Beef cattle feedlots (including veal)	112112	0211
Beef cattle ranching and farming	112111	0212
Hogs	11221	0213
Sheep	1241, 11242	0214
General livestock, except dairy and poultry	11299	0219
Dairy farms	11212	0241
Broiler, fryers, and roaster chickens	11232	0251
Chicken eggs	11231	0252
Turkey and turkey eggs	11233	0253
Poultry hatcheries	11234	0254
Poultry and eggs	11239	0259
Ducks	112390	0259
Horses and other equines	11292	0272

Table 4.2
Summary of CAFO size thresholds for all animal sectors (USEPA 2003a)

Sector	Large	Medium*	Small†
Cattle or cow-calf pairs	1,000 or more	300-999	Less than 300
Mature dairy cattle	700 or more	200-699	Less than 200
Veal calves	1,000 or more	300-999	Less than 300
Swine (weighing over 55 pounds)	2,500 or more	750-2,499	Less than 750
Swine (weighing less than 55 pounds)	10,000 or more	3,000-9,999	Less than 3,000
Horses	500 or more	150-499	Less than 150
Sheep or lambs	10,000 or more	3,000-9,999	Less than 3,000
Turkeys	55,000 or more	16,500-54,999	Less than 16,500
Laying hens or broilers‡	30,000 or more	9,000-29,999	Less than 9,000
Chickens other than laying hens§	125,000 or more	37,500-124,999	Less than 37,500
Laying hens§	82,000 or more	25,000-81,999	Less than 25,000
Ducks§	30,000 or more	10,000-29,999	Less than 10,000
Ducks‡	5,000 or more	1,500-4,999	Less than 1,500

*Must also meet one of two "method of discharge" criteria to be defined as a CAFO or may be designated
†Not a CAFO by regulatory definition, buy may be designated by a CAFO on a case-by-case basis
‡Liquid manure handling systems
§Other than liquid manure handling systems

USEPA expects that the Rule will regulate approximately 15,500 CAFOs that produce close to 60 percent of all manure generated by operations that confine animals (CAFOs represent nearly 60 percent of the 500-million tons of manure produced annually by these operations) (USEPA 2003a). Benefits are expected in terms of both reduced surface water and ground water pollution.

USEPA estimates that for every year the CAFO Rule is implemented, 166 million pounds of nutrients, 3,332 x 10^{19} cfu (colony-forming units) of pathogens and 2.2-billion pounds of sediment loads will be removed from runoff from land where manure has been applied from both medium-size and large-size CAFOs. They also estimate that hydrogen sulfide emissions from large CAFOs will be reduced by 12 percent and methane emissions by 11 percent (USEPA 2003a).

Economic benefits, which USEPA estimates at $204 to $355 million, may include the following (USEPA 2003a):

- Increased use of waters for recreation ($166 to $299 million a year)
- Reduced nitrate contamination of private drinking wells ($31 to $46 million a year)
- Improved shellfish harvests ($0.3 to $3.4 million a year)
- Fewer fish kills ($100,000 a year)
- Cost savings for drinking water treatment ($1.1 to $1.7 million a year)
- Reduced loss of livestock to disease ($5.3 million a year)

Additional benefits that USEPA could not estimate a monetary value for include reduced eutrophication; reduced fecal contamination; reduced human health and ecological risks from pathogens, metals, and salts; improved soil quality; and reduced use of fertilizers.

As part of their assistance for the CAFO rule, USEPA maintains a website with links to state regulations, permits and policies, nutrient management information, livestock and trade associations, federal web sites, best management practices and controls, cooperative extension and land grant universities, research, funding, information on environmental issues, and assorted other information sources (see http://cfpub.epa.gov/npdes/afo/virtualcenter.cfm). Guidelines for NPDES permitting authorities, owners and operators of animal feeding operations, and the general public on how to implement the CAFO regulations can be found in USEPA's NPDES Permit Writers' Guidance Manual and Example NPDES Permit for Concentrated Animal Feeding Operations (USEPA 2003b). USEPA has also published another guide specifically for CAFO operators (USEPA 2003c).

In support of the CAFO Rule, USEPA provided evidence for expected improvements in water quality and reduced drinking water treatment costs related to total suspended solids (Chapter 10 in USEPA 2002a), as well as for reduced nitrate contamination of private wells (Chapter 7 in USEPA 2002a http://www.epa.gov/npdes/pubs/cafo_benefit_p3.pdf).

In addition, USEPA and USDA previously developed a National Unified Strategy for Animal Feeding Operations (USDA 1999) to minimize water quality and public health impacts from AFOs. The goal of the unified strategy is for AFO owners and operators to take actions to minimize water pollution from confinement facilities and land application of manure. The strategy expects that all other AFOs should develop and implement a site-specific, economically feasible, and technically sound Comprehensive Nutrient Management Plan (CNMP). AFOs not defined as CAFOs are encouraged to participate in voluntary programs conducted by federal and state agencies to learn about technical and financial assistance available for developing and implementing a CNMP for their operation.

As necessary, a CNMP should address the following:

- Feed management
- Manure handling and storage
- Land application of manure
- Land and soil conservation practices
- Contaminated runoff from the confinement area
- Proper mortality disposal
- Record keeping

USEPA maintains webpages with the following state-specific information:

- Links to various state environmental agencies responsible for administering AFO and CAFO regulatory programs. The state web sites contain state regulations and requirements for AFOs and CAFOs, downloadable example general and individual CAFO permits, fact sheets, inspection forms, guidance, and compliance assistance publications (http://cfpub.epa.gov/npdes/afo/afoinfo.cfm?view=category&link_cat=24). A second similar list is also available (http://cfpub.epa.gov/npdes/linkresult.cfm?program_id=7&link_category=2&view=link)
- Descriptions of state AFO and CAFO regulations and requirements (http://cfpub.epa.gov/npdes/afo/afoinfo.cfm?view=category&link_cat=23)
- A list of state environmental agency primary personnel contacts for AFO/NPDES issues (http://cfpub.epa.gov/npdes/contacts.cfm?program_id=7&type=STATE)

- For each state, a set of links to state agency AFO web sites and land grant universities (http://cfpub.epa.gov/npdes/afo/virtualcenter.cfm http://cfpub.epa.gov/npdes/afo/virtualcenter.cfm#state_search)
- A report prepared in 2002 for USEPA that provides a compilation of descriptions of state AFO-related regulatory programs and initiatives, with a specific focus on the use of permits or similar mechanisms (Tetra Tech Inc. 2002; www.epa.gov/npdes/pubs/statecom.pdf)

Challenges to the USEPA CAFO Rule

Various components of the CAFO Rule were challenged in court by both agricultural organizations saying the rules exceeded USEPA's authority (American Farm Bureau Federation, National Pork Producers Council, National Chicken Council, and National Turkey Federation, although the National Turkey Federation later withdrew its petition) and environmental groups claiming the rule did not provide enough regulatory control and enforcement over CAFO discharges (Waterkeeper Alliance, Natural Resources Defense Council, Sierra Club, and American Littoral Society). The lawsuits were consolidated, and were decided by the US 2nd Circuit Court of Appeals on February 25, 2005 (Waterkeeper Alliance Inc. et al. v. USEPA, 399 F.3d 486, 2nd Cir. 2005). The complete text of the decision can be found at www.epa.gov/npdes/pubs/cafo_decision.pdf or www.cdqa.org/pdf/decision.pdf, and a summary is available at www.epa.gov/npdes/pubs/summary_court_decision.pdf. A description of the unaffected parts of the CAFO Rule is available at www.epa.gov/npdes/pubs/cafo _minirule_ q_and_a.pdf.

The Court upheld most provisions of the 2003 CAFO Rule but vacated and remanded others. While the Court denied many of the challenges, they agreed that the process used to review Nutrient Management Plans (NMPs) did not provide adequate oversight by state and federal authorities or allow for adequate public input. The court also upheld environmentalists' claim that the rule failed to require CAFOs to use the necessary controls to reduce bacteria and other pathogens, as required by the Clean Water Act. Additionally, the Court upheld farm organizations' claim that the USEPA had gone too far in requiring all CAFOs to apply for discharge permits (based on their "potential to discharge"), regardless of whether or not they actually discharge pollutants to surface waters (i.e., the court decided that a farm must discharge in order for USEPA to have the authority to require NPDES permits). In summary, the Court vacated "those provisions of the CAFO Rule that: (1) allow permitting authorities to issue permits without reviewing the terms of the nutrient management plans; (2) allow permitting authorities to issue permits that do not include the terms of the nutrient management plans and that do not provide for adequate public participation; and (3) require CAFOs to apply for NPDES permits or otherwise demonstrate that they have no potential to discharge." (http://agenvpolicy.aers.psu.edu/Documents/CAFORule2ndCircuitDecisionSummary.pdf).

Permitting requirements for dry-litter poultry facilities in North Carolina are an example of the effect of the Court's decision. All of the dry-litter poultry facilities in North Carolina traditionally fall under what are called "deemed permits", meaning that there are specific regulations that the facilities must meet (mainly NMPs and record-keeping) but there is no formal permitting process. There are also no routine inspections of the facilities. The USEPA CAFO Rule effective in 2003 would have required the larger dry litter facilities to apply for and receive coverage under an NPDES permit. The State of North Carolina was in the process of developing that program when the 2nd Circuit decision was handed down. Based on that decision, the State of North Carolina suspended their NPDES permitting program for dry-litter facilities. All of North Carolina's wet-waste system facilities (swine and dairy) are basically unaffected because there is a state permitting program for those facilities regardless of the federal rules.

On February 6, 2006 USEPA issued a Final Rule extending certain CAFO operator compliance deadlines for the CAFO Rule in order to allow USEPA more time to comply with the Court's decision ("Revised Compliance Dates for National Pollutant Discharge Elimination System Permit Regulation and Effluent Limitation Guidelines for Concentrated Animal Feeding Operations", Federal Register, Vol. 71, No. 28, February 6, 2006; available at http://www.epa.gov/fedrgstr/EPA-WATER/2006/February/Day-10/w1240.htm). This Rule gives livestock operations until July 31, 2007 to develop and implement nutrient management plans and obtain NPDES permits. Under the 2003 rule, CAFOs were to obtain NPDES permits by February 13, 2006, and to develop and implement nutrient management plans to manage animal wastes by December 31, 2006.

The USEPA Administrator signed on June 19, 2006 and issued for public comment a proposed rule to revise the 2003 CAFO regulations more broadly to address the ruling of the Second Circuit Court of Appeals (see http://cfpub.epa.gov/npdes/afo/aforule.cfm). The primary components of this proposed rule are to:

1. Require only the owners and operators of those CAFOs that discharge or propose to discharge to seek coverage under a permit
2. Require CAFOs seeking coverage under a permit to submit their nutrient management plan (NMP) with their application for an individual permit or notice of intent to be authorized under a general permit.
3. Authorize permit writers, upon request by a CAFO, to establish best management, zero discharge effluent limitations when the facility demonstrates that it has designed an open containment system that will comply with the no discharge requirements

WASTE GENERATION AND COMPOSITION FROM AFOs

This section provides a review of the quantity and quality characteristics of different wastes generated at the various types of AFOs (e.g., swine, chicken, turkey, feed cattle, dairy cattle, and aquaculture). Within each AFO type, there are different waste sources, such as manure (feces and urine), spent bedding materials (litter), animal parts, animal mortality, excess feed, feed lot runoff, and others. The four waste categories explicitly regulated by USEPA for facilities designated as CAFOs include manure, litter, process water, and overflow.

The species, size and maturity of the animals, as well as the composition of their feed, affect the quantity and characteristics of the manure and other wastes. The primary pollutants associated with animal wastes are nutrients (primarily nitrogen and phosphorus), pathogens, salts, organic matter, solids, and volatile and odorous compounds. To a lesser extent animal wastes are also a source of trace elements, pharmaceuticals (antibiotics and hormones), and pesticides. AFOs and CAFOs may also have waste containing soaps, surfactants, acids, alkalines, and disinfectants used for cleaning, health maintenance, and food safety. A listing of these contaminants, their primary modes of transport in the environment, and potential impacts on human health and the environment, is provided in Table 4.3. While salts are typically present in waste regardless of animal or feed type, trace elements and pharmaceuticals are usually the result of feed additives used to help prevent disease or promote growth. As such, the type and concentrations of trace elements and pharmaceuticals will vary with operation type and from facility to facility.

Table 4.3

Key pollutants in animal waste

(Derived from Exhibit 2-2 in USEPA 2002a)

Pollutant	Description of pollutant forms in animal waste	Pathways	Potential impacts
Nitrogen	Exists in fresh manure in organic (e.g., urea) and inorganic forms (e.g., ammonium and nitrate). Microbes transform organic nitrogen to inorganic forms that may be absorbed by plants	▪ Overland discharge ▪ Leachate into ground water ▪ Atmospheric deposition as ammonia	▪ Eutrophication ▪ Animal, human health effects ▪ Groundwater degradation
Phosphorus	Exists in both organic and inorganic forms. As manure ages, phosphorus mineralizes to inorganic phosphate compounds that may be absorbed by plants.	▪ Overland discharge ▪ Leachate into groundwater (water soluble forms)	▪ Eutrophication ▪ Surface water degradation
Potassium	Most potassium in manure is in an inorganic form available for absorption by plants; it can also be stored in soil for future uptake.	▪ Overland discharge ▪ Leachate into groundwater	▪ Increased salinity
Organic compounds	Carbon-based compounds in manure that are decomposed by soil and surface water micro-organisms. Creates biochemical oxygen demand, or BOD, because decomposition consumes dissolved oxygen in the water. Degradation of organic compounds (mineralization) can create salts, primarily bicarbonate, and nutrients.	▪ Overland discharge	▪ Depletion of dissolved oxygen ▪ Reduction in aquatic life ▪ Eutrophication ▪ Groundwater degradation
Solids	Includes manure itself and other elements (e.g., feed, bedding, hair, feathers, and corpses).	▪ Overland discharge ▪ Atmospheric deposition	▪ Turbidity ▪ Siltation
Pathogens	Includes range of disease-causing organisms, including bacteria, viruses, protozoa, fungi, and algae. Some pathogens are found in manure, others grow in surface water due to increased nutrients and organic matter.	▪ Overland discharge ▪ Growth in waters with high nutrient, organic materials	▪ Animal, human health effects ▪ Groundwater degradation
Salts	Includes cations sodium, potassium, calcium, and magnesium; and anions chloride, sulfate, bicarbonate, carbonate, and nitrate.	▪ Overland discharge ▪ Leachate into groundwater	▪ Reduction in aquatic life ▪ Human health effects ▪ Increased salinity ▪ Soil impacts
Trace elements	Includes feed additives arsenic, copper, selenium, zinc, cadmium; and trace metals molybdenum, nickel, lead, iron, manganese, aluminum, and boron (pesticide ingredients)	▪ Overland discharge	▪ Toxicity at high levels
Volatile compounds	Includes carbon dioxide, methane, nitrous oxide, hydrogen sulfide, and ammonia gases generated during decomposition of waste.	▪ Inhalation ▪ Atmospheric deposition of ammonia	▪ Human health effects ▪ Eutrophication ▪ Global warming ▪ Groundwater degradation
Other pollutants	Includes pesticides, antibiotics, hormones, soaps and other surfactants, acids and alkaline, and disinfectants (used for cleaning, health maintenance, and food safety) used in feeding operations	▪ Overland discharge	▪ Impacts unknown ▪ Groundwater degradation

For surface waters, the most common problems for water suppliers associated with AFO wastes in their source water area are excessive nutrients (especially phosphorus, and also nitrogen) causing eutrophication and resulting increases in organic matter, and taste and odor, as well as the presence of microbial pathogens such as *Cryptosporidium*. For groundwater supplies, nitrate and pathogens have generally caused the most problems.

Agricultural waste may be in the form of a liquid, slurry, semi-solid, solid, or gas. Percent moisture, total solids, nutrient content, nutrient composition (i.e., C:N ratio), and other waste characteristics can impact handling and treatment, including land application. As such, wastes of different consistencies may require different management techniques and handling equipment (USDA 1999).

Manure wastes can change consistency throughout the handling systems or throughout the year. The primary characteristic that determines how the waste may physically be handled is its total solids concentration. The type of animal, amount of water consumed, feed type and climate can all affect the solids concentration of excreted manure. In most cases the consistency of the waste can be anticipated or determined. The total solids concentration of the waste can be increased by adding solids (e.g., bedding) to the waste, decreased by adding water, or stabilized (by protecting it from absorbing additional water). Figure 4.7 illustrates the consistency of manure for different animals for various total solids concentrations (USDA 1999).

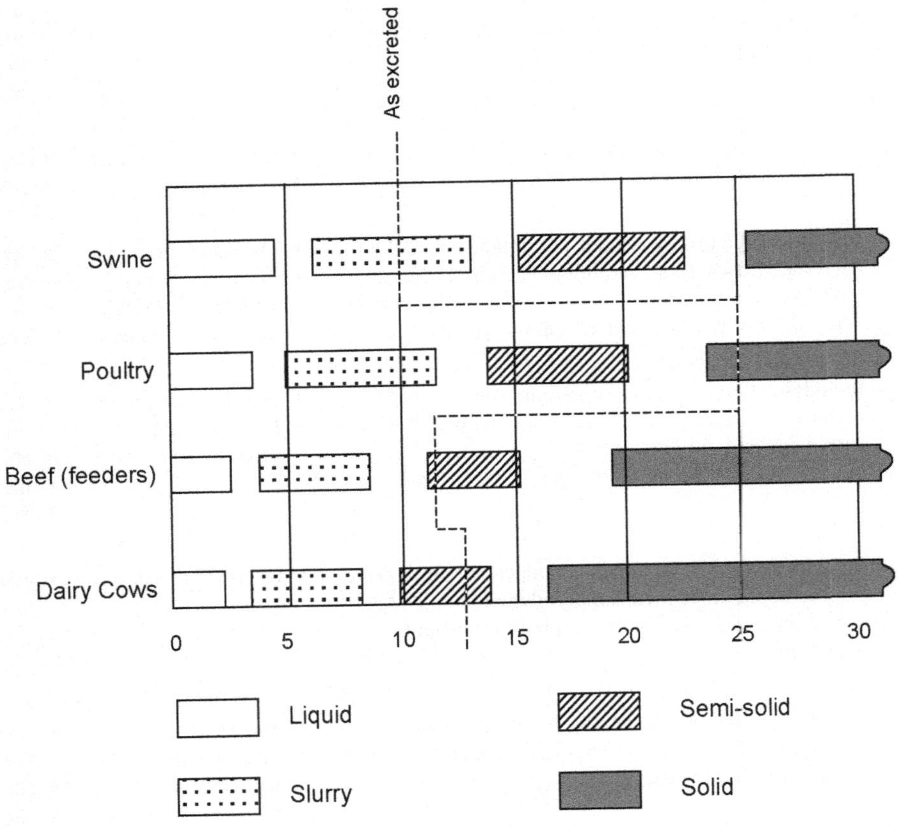

Source: USDA (1999)

Figure 4.7 Relative handling characteristics of different kinds of manure and percent total solids

The tables that follow show typical waste characteristics for a variety of animals and waste types. Table 4.4a shows estimated characteristics of manure production (urine and feces combined) from meat-producing livestock and poultry, including the mass and volume of manure produced and typical concentrations of key parameters. Table 4.4b shows the same for other livestock and poultry. These data are for the wastes as excreted by the various animals. Table 4.5 provides description of typical characteristics of manure after it has undergone various handling procedures (i.e., 'as-removed'). Manure characteristics can be altered by many physical, chemical and biological processes. Common handling techniques shown in Table 4.5 include water addition or removal, bedding addition, and treatment processes. The variation of as-removed waste is substantial and correlated to the type of manure management system in use as well as the geographic location (ASAE 2005). Numerous other tables with slightly different waste characteristic information are available for different animal types in Chapter 4 of the USDA's Agricultural Waste Management Field Handbook (USDA 1999).

AFO-DERIVED CONTAMINANTS IN THE ENVIRONMENT

Animal wastes contain many constituents that can adversely affect humans and water quality. Release of these wastes to the environment can have deleterious effects on both groundwater and surface water quality. USEPA (2002) concluded that potential human health impacts from manure-related waterborne contaminants are primarily associated with drinking contaminated water, direct contact with contaminated water, and consuming contaminated shellfish. Release to the environment of trace metals as well as gases from AFO operations are also of concern.

Nutrients in animal wastes can cause oxygen depletion and eutrophication in surface waters, with a potential for algal blooms and fish kills. Nitrogen contamination of groundwater can lead to elevated levels of nitrate in drinking water wells. Pathogenic microorganisms are also of significant concern, especially for water systems that have no disinfection or for those microorganisms that are resistant to the disinfectant(s) used. Solids from animal waste can increase turbidity and adversely affect the taste and odor of waters. In addition, metals such as arsenic, copper, selenium, and zinc, which are often added to animal feed (and hence are excreted in manure), can be toxic to humans in sufficient concentrations over time. Antibiotics, hormones and pesticides, also used in animal feeding operations, can also be of concern.

Numerous studies have indicated that animal manure can be a significant source of pollutants that contributes to water quality degradation. A literature survey conducted in preparation for development of the USEPA CAFO rule identified more than 150 reports of discharges to surface waters from hog, poultry, dairy, and cattle operations, including over 30 separate incidents of swine operation discharges in Iowa alone between the years 1992 and 1997. The waste discharge incidents resulted in fish kills ranging from about 500- to more than 500,000-fish killed per event (USEPA 2002a, 2003a).

Table 4.4a

Estimated typical manure characteristics for meat-producing livestock and poultry (as-excreted)

(Derived from ASAE 2005: Standard D384.2, Manure Production and Characteristics)

Animal type and production grouping	Total solids[3]	Volatile solids[3]	COD[3.4]	BOD[3.4]	Nitrogen	P	K	Ca	Total manure[5]		Moisture[6]
	kg/finished animal (f.a.)								kg/f.a.	liter/f.a.	%w.b.
Beef – Finishing cattle	**360**	**290**	300	67	**25**	**3.3**	17.1	**7.7**	4,500	4,500	92
Poultry – Broiler	**1.3**	0.95	1.05	0.30	**0.053**	**0.016**	**0.031**		4.9	4.9	74
Poultry – Turkey (male)	**9.2**	7.4	8.5	2.4	**0.55**	**0.16**	0.26		36	36	74
Poultry – Turkey (female)	**4.4**	3.5	4.0	1.1	**0.26**	**0.074**	0.11		17	17	74
Poultry – Duck	**1.7**	1.0	1.4	0.28	**0.062**	**0.022**	0.031		6.5	6.5	74
Swine – Nursery pig (12.5 kg)	**4.8**	4.0	4.4	1.5	**0.41**	**0.068**	0.16		48	48	90
Swine – Grow-finish (70 kg)	**56**	45	47	17	**4.7**	**0.76**	2.0		560	560	90
	lb/finished animal (f.a)								lb/f.a.	ft^3/f.a.	%w.b.
Beef – Finishing cattle	**780**	**640**	670	150	**55**	**7.3**	38	**17**	9,800	160	92
Poultry – Broiler	**2.8**	2.1	2.3	0.66	**0.12**	**0.035**	**0.068**		11	0.17	74
Poultry – Turkey (male)	**20**	16	19	5.2	**1.2**	**0.36**	0.57		78	1.3	74
Poultry – Turkey (female)	**9.8**	7.8	8.8	2.4	**0.57**	**0.16**	0.25		38	0.61	74
Poultry – Duck	**3.7**	2.2	3.0	0.61	**0.14**	**0.048**	0.068		14	0.23	74
Swine – Nursery pig (27.5 lb)	**10**	8.7	9.7	3.4	**0.91**	**0.15**	0.35		87	1.4	90
Swine – Grow-finish (154 lb)	**120**	99	104	38	**10**	**1.7**	4.4		1200	20	90

Note: For footnotes, see Table 4.4b
Diet based numbers are in BOLD. See footnotes 2 and 3 in Table 4.4b for source of non-bold values

Table 4.4b

Estimated typical manure characteristics for other livestock and poultry (as-excreted)

(Derived from ASAE 2005: Standard D384.2, Manure Production and Characteristics)

Animal type and production grouping	Total solids[3]	Volatile solids[3]	COD[3,4]	BOD[3,4]	Nitrogen	P	K	Ca	Total manure[5]		Moisture[6]
	kg/day-animal (d-a)								kg/(d-a)	liter/(d-a)	%w.b.
Beef – Cow (confinement) [7,10]	**6.6**	**5.9**	6.2	1.4	**0.19**	**0.044**	0.14	**0.089**	-	-	88
Beef – Growing calf (confinement)	**2.7**	**2.3**	2.3	0.52	**0.13**	**0.025**	0.085	**0.040**	22	22	88
Dairy – Lactating cow	**8.9**	7.5	8.1	1.30	**0.45**	**0.078**	**0.103**		68	68	87
Dairy – Dry cow	**4.9**	4.2	4.4	0.626	**0.23**	0.03	0.148		38	3	87
Dairy – Milk fed cows					0.0079						
Dairy – Calf-150 kg	1.4				0.063				8.5	8.5	83
Dairy – Heifer-440 kg	3.7	3.2	3.4	0.54	**0.12**	**0.020**			22	22	83
Dairy – Veal-118 kg	0.12				0.015	0.0045	0.0199		3.5	3.5	96
Horse – Sedentary-500 kg[8]	**3.8**	3.0		0.48	**0.089**	**0.013**	**0.027**	**0.023**	25	25	85
Horse – Intense exercise-500 kg[8]	**3.9**	3.1		0.49	**0.15**	**0.033**	**0.095**	**0.069**	26	26	85
Layer	**0.022**	0.016	0.018	0.0050	**0.0016**	**0.00048**	0.00058	**0.0022**	0.088	0.088	75
Swine – Gestating sow-200 kg	0.50	0.45	0.47	0.17	**0.032**	**0.009**	0.022		5.0	5.0	90
Swine – Lactating sow[9]-192 kg	1.2	1.0	1.1	0.38	**0.085**	**0.025**	.053		12	12	90
Swine – Boar-200 kg	0.38	0.34	0.27	0.13	0.028	0.0097	.0176		3.8	3.8	90
	lb/day-animal (d-a)								lb/(d-a)	ft^3/(d-a)	%w.b.
Beef – Cow (confinement)[7,10]	**15**	13	14	3.0	**0.42**	**0.097**	0.30	**0.20**	-	-	88
Beef – Growing calf (confinement)	**6.0**	5.0	5.2	1.1	**0.29**	**0.055**	0.19	**0.088**	50	0.81	88
Dairy – Lactating cow	**20**	17	18	2.9	**0.99**	**0.17**	**0.23**		150	2.4	87
Dairy – Dry cow	**11**	9.2	9.7	1.4	**0.50**	**0.066**	0.33		83	1.3	87
Dairy – Milk fed calves					0.017						
Dairy – Calf-330 lb	3.2				0.14				19	0.30	83
Dairy – Heifer-970 lb	8.2	7.1	7.5	1.2	**0.26**	**0.044**			48	0.78	83
Dairy – Veal-260 lb	0.27				0.033	0.0099	0.044		7.8	0.12	96

(continued)

Table 4.4b (Continued)

Animal type and production grouping	Total solids[3]	Volatile solids[3]	COD[3.4]	BOD[3.4]	Nitrogen	P	K	Ca	Total manure[5]		Moisture[6]
				lb/day-animal (d-a)					lb/(d-a)	ft^3/(d-a)	%w.b.
Horse – Sedentary-1,100 lb[8]	**8.4**	6.6		1.1	**0.20**	**0.029**	**0.060**	**0.051**	56	0.90	85
Horse – Intense exercise-1,100 lb[8]	**8.6**	6.8		1.1	**0.34**	**0.073**	**0.21**	**0.15**	57	0.92	85
Layer	**0.049**	0.036	0.039	0.011	**0.0035**	**0.0011**	0.0013	0.0048	0.19	0.0031	75
Swine – Gestating sow-440 lb	1.1	0.99	1.0	0.37	**0.071**	**0.020**	0.048		11	0.18	90
Swine – Lactating sow[9]-423 lb	2.5	2.3	2.4	0.84	**0.19**	**0.055**	0.12		25	0.41	90
Swine – Boar-440 lb	0.84	0.75	0.60	0.29	0.061	0.021	0.039		8.4	0.13	90

Notes: Diet based numbers are in **BOLD**. See footnotes 2 and 3 below for source of non-bold values

1. Prior to any changes due to dilution water addition, drying, volatilization or other physical, chemical, or biological processes
2. Non-bold table numbers indicate that predictive equations were not available from Sections 4-9 for estimating this characteristic. These numbers are average values taken from MWPS-18 Section 1, NRCS Agricultural Waste Management Field Handbook, and the previous version ASAE D384.1 or calculated based upon procedures used in footnote 3
3. Total Solid (TS) is estimated for most animal groups by equations in Sections 4-9. For beef cattle, volatile solids is also based upon equations. For all species, volatile solids are calculated from TS and literature values of the ratio of VS to TS. Similarly, BOD and COD values are calculated using VS and the literature values of the ratio of BOD and COD to VS. Literature values are taken from MWPS-18 Section 1, NRCS Agricultural Waste Management Field Handbook, and the previous version ASAE D384-1
4. BOD – Biochemical Oxygen Demand (5-day), COD – Chemical Oxygen Demand
5. Total manure is calculated from Total Solids and manure moisture content
6. As-excreted manure moisture contents range from 75 to 90 percent. At these moisture levels as-excreted manure has a density nearly equal to that of water, and a specific gravity of 1.0 was assumed in calculation of manure volume
7. Solids estimates (TS, VS, COD, and BOD) do not include solids in urine
8. These values apply to horses 18 months of age or older that are not pregnant of lactating. The representative number applies to 500 kg horses and the range represents horses from 400 to 600 kg. "Sedentary" would apply to horses not receiving any imposed exercise. Dietary inputs are based on minimum nutrient requirements specified in "Nutrients Requirements of Horses" (NRC 1989). "Intense" represents horses used for competitive activities such as racing. Dietary inputs are based on a survey of race horse feeding practices (Gallagher et al. 1992) and typical feed compositions (forage = 50% alfalfa, 50% timothy; concentrate = 30% oats, 70% mixed performance horse concentrate)
9. Bold values include contribution of nursing pigs
10. Beef cow values are representative of animals during non-lactating period and first six months of gestation

Table 4.5
Estimated typical manure characteristics (as-removed)
(Derived from ASAE 2005: Standard D384.2, Manure Production and Characteristics)

Description	Moisture (percent)	Total solids (percent)	Volatile solids (% of TS)	TKN	NH_3-N	P	K	Ca	Na	Zn	Mn	Cu	Mass (kg/hd/d)
				Percent (wet basis)					ppm (wet basis)				
Beef													
Earthen lot	33	67	30	1.18	0.10	0.50	1.25	1.21	3,012	85	393	14	7.5
Poultry													
Leghorn pullets	65	40		2.13	0.85	1.00	1.10	1.49					No data
Leghorn hen	59	40		1.85	0.88	1.21	1.31	6.40					0.03
Broiler litter	31	70	70	3.73	0.75	0.60	1.37	1.82					0.02
Turkey litter	30			2.18		0.33	1.23	5.00					0.11
Dairy													
Scraped earthen lots	54	46		0.70		0.25	0.67	0.45	311	1.2	1.4	0.2	35
Scraped concrete lots	72	25		0.53		0.13	0.40	0.31	32	0.4	0.7	0.1	40
Lagoon effluent	98	2	52	0.073	0.08	0.016	0.11	0.04	7	0.9	1.4	2	106
Slurry (liquid)	92	8	66	0.30	0.14	0.13	0.40	0.40	905	25	40	75	67
Equine													
Solid manure	43	65	26	0.76		0.24	0.99	1.13		51.7	135	12.70	*
Swine													
Finisher-slurry wet-dry feeders	91	9		0.70	0.50	0.21	0.24	0.25	380	85.0		50	3-4
Slurry storage-dry feeders	94	6		0.47	0.34	0.18	0.24	0.25	380	68		30	4.5
Flush building	98	2		0.20	0.14	0.07	0.17	0.04	300	33.6	14.4	31.2	16
Agitated solids & water	98	2		0.10	0.05	0.06	0.06	0.08	215	44.4	15.6	19.2	
Lagoon surface water	99.6	0.40		0.06	0.04	0.02	0.07	0.01	215	3.6	1.2	2.4	
Lagoon sludge	90	10		0.26	0.07	0.25	0.07	0.04	191	22	80	90	

*32.2 kg per head per day for residential, 46 kg per head per day for commercial

An understanding of the principles of the fate and transport of animal waste contaminants in the environment is important for water utilities to be able to more accurately assess potential impacts on source water quality from releases of these wastes. Wastes can be released to the environment during any of the animal waste generation and handling steps, including the animal production areas, treatment and storage lagoons, manure stockpiles, and land from where the manure is applied. The main transport pathways include surface runoff and erosion, overflows from liquid waste storage facilities (e.g., lagoons and storage ponds), spills and other dry-weather discharges, leaching into soil and ground water, and volatilization of compounds (e.g., ammonia) and subsequent redeposition on land or surface water (USEPA 2003a). Preferential flow paths can be especially important pathways for subsurface transport of contaminants.

Some releases to the environment may be continuous (e.g., gases volatizing from liquid manure lagoons), some releases may be periodic (e.g., nutrient release to ground water or surface water after land application of manure), and others may be short-term and more substantial (e.g., release of large volumes of liquid manure waste from a lagoon with a breached berm).

Various factors affect the fate and transport of contaminants from waste sources to the subsurface or into surface water sources, including hydrogeologic and weather conditions, and the physical and chemical properties of the contaminant. Fate and transport mechanisms include the following:

- Water transport related processes: (a) advection, (b) dispersion, (c) diffusion
- Phase transfer processes: (a) immobilization, (b) sorption, (c) ion exchange, (d) volatilization, (e) dissolution, and (f) precipitation
- Species transformation processes: (a) reduction/oxidation reactions, (b) acid-base reactions, (c) substitution reactions, (d) hydrolysis (e) complexation, (f) biological degradation, (g) photochemical degradation, and (h) radioactive decay

Runoff of animal wastes applied to land is more likely when rainfall occurs shortly after application. This is especially true if the manure is overapplied, misapplied, or is not injected or incorporated into the land. Furthermore, erosion of soil that has contaminants (e.g., phosphorus, pathogens) bound to it can also occur during rain events. More direct pollution of surface waters occurs when animals are allowed direct access to waterbodies (e.g., cows allowed to drink in streams); or when manure storage areas leak, overflow, or their containment system is breached; or when manure or other piping systems leak or break. Flooding from severe precipitation events can result in substantial release of wastes from a variety of areas. Animal wastes can also reach groundwater or surface water during dry periods, resulting from overflow of manure storage areas, spills when containment systems fail, or intentional releases (e.g., in irrigation water). Contamination of groundwater can occur through many of the same sources, and is especially predominant when manure or other pollutants are applied to land, and then rainfall or excess irrigation carries the contaminants into the subsurface (e.g., nitrate from animal wastes).

Contamination of air is also a possibility through release of volatile compounds from waste or byproducts of animal waste decomposition such as ammonia, hydrogen sulfide and methane. Spray irrigation systems and wind-blown dust particles also are a source of atmospheric contamination. Any chemical that becomes airborne may then later be redeposited on surface waters or soils (USEPA 2003a).

The factors affecting the transport of AFO-derived contaminants, as well as potential impacts on source water quality, human health and the environment, are discussed below for several parameters of potential concern.

Pathogens

A significant health and regulatory concern associated with livestock, poultry, and horse wastes are pathogens, particularly organisms like *Cryptosporidium* that are relatively resistant to free chlorine and other common drinking water disinfection practices. Pathogens are defined as disease-causing organisms, typically microorganisms. Many pathogens found in animal waste can infect humans if ingested. Pathogens of concern to humans include bacteria, viruses, and parasites. More than 150 pathogens have been found in livestock manure that are associated with risks to humans. Among these, the six pathogens that account for more than 90 percent of food and waterborne diseases in humans are *Campylobacter spp., Salmonella spp.* (non-typhoid), *Listeria monocytogenes, Escherichia coli* O157:H7, *Cryptosporidium parvum,* and *Giardia lamblia.* Each of these organisms may be transmitted from one animal to another at an AFO. These organisms also have a relatively low infectious dose in humans (USEPA 2002a, 2003a).

Organisms like *Cryptosporidium*, *Giardia*, and *Salmonella* are frequently found in animal manure and can induce symptoms ranging from skin sores to chest pain. *E. coli*, which causes diarrhea and abdominal gas, has been the source of disease outbreaks in several states. Particularly virulent strains of *E. coli* can cause serious illness and even death. *Cryptosporidium* and *Giardia* are of particular concern because they are highly resistant to disinfection with chlorine, and low doses of either pathogen can cause disease. For example, *Cryptosporidium* causes gastrointestinal illness that lasts 2 to 10 days in healthy individuals but can be fatal in people with weakened immune systems. The bacteria *Listeria monocytogenes* is ubiquitous in nature and is commonly found in the intestines of wild and domestic animals (USEPA 2003a). In addition to the risks of pathogenic contamination of drinking water supplies, animal manure in surface waters can also threaten human health through recreational contact and through exposure to contaminated shellfish (USEPA 2002a, 2003a).

Attenuation and transportation mechanisms for microbial contaminants are more complex than for other contaminants. Furthermore, the most complex aspects of transport of non-microbial contaminants are typically those processes that are microbiologically mediated.

Some pathogens (e.g., *Cryptosporidium parvum* oocysts) can survive in the environment for long periods of time. Several variables and factors control the survival and transport of microbial pathogens from waste application sites into the environment. The factors that can decrease the survival of *Cryptosporidium* oocysts include freeze-thaw cycling, desiccation, elevated temperatures (such as during composting), sunlight (including UV light components), pH (both low and high), and the presence of ammonia (e.g., from urea degradation).

A critical review of available research on transport of pathogens in watersheds was conducted by Ferguson et al. (2003) and Davies et al. (2005). The latter report also presented results of a series of research experiments designed to identify the key factors controlling pathogen transport (Davies et al. 2005). Among the many findings summarized is that temperature is a very influential factor on the survival of *Cryptosporidium* oocysts. Both high temperatures and freezing temperatures can result in *Cryptosporidium* inactivation, while at more moderate temperatures (e.g., 4 to 25°C) inactivation rates are relatively slow (Davies et al. 2005, Finstein 2004). Freeze-thaw cycling is more effective than freezing at reducing survival (or at inactivation), perhaps because of increased mechanical damage to the oocyst wall during the temperature fluctuations (Walker, Leddy, and Hagar 2001). Soil texture also can significantly affect inactivation of *Cryptosporidium* oocysts and viruses (CRC 2004). Vegetative filter strips can be effective at removing *Cryptosporidium* oocysts from surface runoff, however, viruses (e.g., PRD1 bacteriophage) and to a lesser extent bacteria (*E. coli*) are more easily transported than are the larger *Cryptosporidium* oocysts (Davies et al. 2005, CRC 2004). A summary of those findings is also available (CRC 2004). John and Rose (2005) provide a review of the factors affecting microbial survival in groundwater.

Pyke et al. (2003) summarized the occurrence, sources, and fate of *Cryptosporidium* in agricultural environments. Kaplan et al. (2002) generated a database on pathogen occurrence in watersheds ranging from forested to highly urbanized. Approaches for reducing overall pathogen loading within an agricultural watershed are discussed by Rosen et al. (2001). A summary of the fate of *Cryptosporidium* and the effectiveness for *Cryptosporidium* inactivation via different BMPs (e.g., composting, anaerobic digestion, manure slurry storage) was provided by Finstein (2004).

Curriero et al. (2001) demonstrated a correlation between the occurrence of rainfall and waterborne disease outbreaks in the US. This epidemiological linkage lends support to the view that water flow is one of the most important parameters affecting transport of microbial contaminants. Rainfall and rainfall intensity are important factors in the release and detachment of pathogens from fecal matter and their mobilization into surface and groundwater. LeChevallier et al. (1998) and Atherholt et al. (1998) demonstrated a correlation between precipitation events and the concentration of waterborne protozoan pathogens and related indicator parameters (e.g., turbidity) in the Delaware River watershed. Microorganisms can potentially be rapidly transported to surface waters by overland flow (Tyrrel and Quiton 2003). Overland flow transport of microorganisms includes three stages: detachment, transport and deposition.

One recently completed research project (Sturdevant Rees et al., N.d.) investigated the variability of pathogen occurrence and transport through watersheds. Even small rainfall events (less than 0.25 inches) were capable of washing *Giardia* cysts and *Cryptosporidium* oocysts into streams, and saturated or near-saturated ground conditions, and events characterized by rain falling on snow, resulted in higher rates of detection. In addition, little correlation was found between rainfall event accumulation and *Cryptosporidium* or *Giardia* detection. Rather, antecedent rainfall and/or remotely sensed soil moisture data indicating saturated or near saturated conditions may be important for identifying rainfall events where sampling for *Cryptosporidium* and *Giardia* is warranted (Sturdevant Rees et al., N.d.).

Mawdsley et al. (1996a, 1996b) examined runoff and infiltration from vegetated microplots using oocyst-inoculated cow waste slurries. Both horizontal and vertical transport of *Cryptosporidium parvum* oocysts was demonstrated. Although it is generally believed that movement of pathogens through the soil is minimal, recent research has shown that appreciable numbers of *Cryptosporidium* may be transported via preferential flow (macropore flow and fingered flow) from the soil surface through the vadose zone to groundwater under a "worst-case" scenario or extreme events (Darnault 2000; Darnault et al. 2003, 2004).

In an AwwaRF research project, Amirtharajah et al. (2002) investigated the transport of a *Cryptosporidium* surrogate (polystyrene microspheres) through unsaturated soils at an undisturbed field site used for cattle production. Results showed that the vertical migration of polystyrene microspheres in column studies suggests that migration of *Cryptosporidium parvum* oocysts through fine-textured soils is likely to be minimal, and that a small number of these surrogate particles travel through preferential flow paths at field sites, especially after rainfall events (Amirtharajah et al. 2002).

Cryptosporidium is thought to be easily transported over land. Because oocysts are approximately the size of clay/silt particles, the amount of kinetic energy needed to entrain and suspend oocysts in overland flow may be quite small. Overland flow transport models for pathogens were reviewed as part of a USEPA workshop on animal feeding operations (GeoLogics Corp. 2004). The effects of land slopes, vegetation, and rainfall intensities on overland and near-surface transport of *Cryptosporidium parvum* oocysts were examined by Trask et al. (2004). Zhang and coworkers (2001) describe the development and implementation of a model for simulating removal of *Cryptosporidium parvum* oocysts from overland flow, including modeling transport of oocysts through vegetative filter strips.

Nutrients

Animal wastes contain significant quantities of nutrients, particularly nitrogen and phosphorus. Nitrogen and phosphorus can enter the environment through runoff, soil infiltration, and percolation into groundwater, and ammonia can volatize with subsequent redeposition on surface waters and land. USEPA considers nutrients to be the leading cause of impaired lakes, ponds, and reservoirs in the US, the fifth leading cause for impaired rivers and streams, and the second leading cause of impairment reported for the Great Lakes (USEPA 2003).

Nitrate and ammonium may leach into groundwater from litter and manure field applications and liquid waste storage facilities. Nitrate is much more mobile than phosphorus in soil. Nitrate passes readily through soils and can contaminate groundwater while phosphorus binds to soil particles until the phosphorus content of soils build up, leading to phosphorus surface export and subsurface loss (Sharpley et al. 1999, Mallin and Cahoon 2003).

Nitrogen in manure occurs in different forms, including ammonia, nitrate, and organic forms such as amines and proteins. Ammonia and nitrate are valuable as fertilizers for crop growth, but both also have potential adverse environmental effects. Nitrate is particularly valuable as a fertilizer because it is biologically available to plants. Ammonia can be toxic to aquatic life, and exerts a biochemical oxygen demand (BOD) on the receiving water (resulting in reduced oxygen levels), and excessive amounts can lead to eutrophication. Nutrients originating from livestock and poultry operations in the Mississippi River Basin have been identified as contributing to a very large hypoxic zone in the northern Gulf of Mexico (USEPA 2003a). On a positive side, ammonia also serves as a neutralizer for atmospheric acidic gases that would precipitate as acid rain.

Nitrate can cause contamination of both groundwater and surface water. Nitrogen contamination of groundwater is of particular concern, as the nitrogen can easily convert to nitrate as it passes through the vadose (unsaturated) zone. Consumption of nitrate can cause methemoglobinemia (blue baby syndrome) in infants, which reduces the ability of the blood to carry oxygen. If left untreated, methemoglobinemia can be fatal. Because of this health risk, the USEPA drinking water maximum contaminant level (MCL) for nitrate was set at 10 mg/L as N. Nitrate can also cause spontaneous abortions and increased incidence of stomach and esophageal cancers (USEPA 2003a).

USEPA expects that the CAFO rule will reduce nitrate levels in private drinking wells by reducing the amount of manure spread on cropland, thus reducing the amount of contaminants that leach through soils and reach ground water (USEPA 2003a). Nitrate is not removed by conventional drinking water treatment processes, and people using domestic wells are typically at greater risk because these wells are often shallower than municipal wells and are not subject to treatment or wellhead protection requirements.

Some public water systems in agricultural areas use expensive treatment processes to remove excess levels of nitrate from their source water (both surface and ground waters). For example, the Des Moines Water Works (Iowa) built the largest nitrate removal system in the world in 1991 at a cost of nearly four million dollars. Operating costs are approximately $3,000 per day. Des Moines obtains their source water from the Raccoon and Des Moines rivers and from wells close to the Raccoon. Typically the sources are switched to use whichever one has less nitrate. When necessary, the nitrate-removal system is operated to meet the MCL of 10 mg/L (Beeman 2002).

Though phosphorus on rare occasions can directly impact drinking water quality (e.g., phosphorus concentrations greater than 1.0 mg/L can interfere with the coagulation process in drinking water treatment plants (USEPA 2003a)), most of the potential impacts of phosphorus on drinking water quality are due to indirect effects. Water-quality problems associated with phosphorus are generally confined to surface waters. Phosphorus is a necessary nutrient for plant growth, and is usually a limiting factor in growth of aquatic biota. Excessive phosphorus can lead to eutrophication, algal blooms, fish kills, reduced biodiversity, objectionable tastes and odors, growth of toxic organisms (e.g., toxins from blue-green algae), and resulting increased drinking water treatment costs. Excessive algal growth also causes an increase in the organic matter content of source waters, and thus can lead to increased concentrations of potentially harmful disinfection byproducts (e.g., trihalomethanes from chlorination).

Most phosphorus in soils exists as inorganic minerals, generally of low solubility and thus limited potential for leaching into groundwater. Soluble phosphorus is readily available for plant uptake. Inorganic phosphorus can adhere to and move with soil particles during runoff erosion, and thus is a greater threat to surface water quality when land has been recently manured or fertilized before a storm event. Soluble phosphorus can also transport with runoff.

Historically manures were land applied to meet the nitrogen needs of crops (with extra applied to account for issues of availability), and in many cases that led to significant overapplication of phosphorus relative to plant needs (typically by a factor of 5). The net result has been overaccumulation ("banking") of phosphorus in soils, and many soils in regions with significant animal production are now saturated with phosphorus. Today, many new nutrient management plans are based on phosphorus and not nitrogen.

Good summaries of information about the source, fate, and transport of phosphorus and nitrogen in the agricultural environment, and applicable BMPs, are provided in the following sources:

- Nitrogen and phosphorus in the environment (Heartland Regional Water Coordination Initiative): www.oznet.ksu.edu/waterquality/welcome.htm
- Nitrogen Fate and Transport (USDA-NRCS): www.nrcs.usda.gov/technical/land/pubs/wp07text.html
- Phosphorus Fate and Transport (USDA-NRCS): www.nrcs.usda.gov/technical/land/pubs/wp08text.html
- BMPs for Phosphorus in the Environment (Walker 2000): www.utextension.utk.edu/publications/pbfiles/PB1645%20.pdf
- Phosphorus, Agriculture & The Environment (Mullins 2001): www.ext.vt.edu/pubs/grains/424-029/424-029.html

Pharmaceuticals (Antibiotics and Hormones)

Veterinary pharmaceuticals can be used to promote growth, fight disease, or provide other therapeutic goals. As an example, antibiotics can be used to treat illness and as feed additives to promote growth or improve feed conversion efficiency. Drugs that fight disease act by interfering with specific receptors, enzymes, or other biological processes, and can potentially remain biologically active when released from animal wastes.

USEPA estimates that between 60 and 80 percent of all livestock and poultry receive some type of antibiotic during their life (USEPA 2003a). Almost all antibiotics and other pharmaceuticals given are eventually excreted in urine and bile either unchanged or in metabolite form. The release of antibiotics to the environment may contribute to the emergence of strains of pathogens that are more resistant to antibiotics (USEPA 2003a). Drug residues in animal food products are also of potential concern for adverse impacts on human health. Interest in veterinary pharmaceuticals as potential emerging contaminants has been sparked in recent years by the proliferation of large-scale AFOs. USEPA has provided a brief overview of pharmaceutical monitoring methods and pharmaceutical fate and transport models (GeoLogics Corporation 2004).

A study by the USGS detected a variety of pharmaceuticals and hormones in US surface waters downstream of areas with intense livestock production. However, the measured concentrations were generally low and rarely exceeded drinking water guidelines, health advisories, or aquatic-life criteria (Kolpin et al. 2002). In a USGS study of liquid waste at swine AFOs in Iowa and North Carolina, multiple classes of antibiotics were detected ranging from ppb to ppm concentrations. Compilation of analytical data for liquid waste from swine operations between 1998 and 2002 found one or more antibiotics present in all of the samples. Antibiotics also have been detected in poultry litter from the Delaware-Maryland-Virginia peninsula. The data from these studies demonstrate that veterinary pharmaceuticals are excreted and frequently occur at detectable levels ranging from ppb to ppm concentrations in liquid and solid waste (http://toxics.usgs.gov/regional/emc/animal_feeding.html). A USGS website is available that provides additional information on pharmaceuticals and other emerging contaminants in the environment (http://toxics.usgs.gov/regional/emc/index.html).

The primary sources of hormones in surface waters are disposal of effluents from municipal wastewater treatment plants and animal feeding operations. Hormones in animal waste can include both naturally-occurring hormones and synthetic hormones. Synthetic hormones are used to increase growth and productivity in the dairy, beef, poultry, and aquaculture industries. Naturally-produced steroid hormones such as androgens, estrogens, and progestins have been detected in wastewater samples from dairy farms and aquaculture facilities, as steroid hormones and their metabolites are excreted by all vertebrates (Kolodziej, Harter, and Sedlak 2004). Hanselman, Graetz, and Wilkie (2003) provided a review of the occurrence of estrogenic hormones in livestock waste, their physicochemical properties, and fate in the environment. Growth hormones given to animals can be extremely short-lived and chemically unstable. For example, synthetic Bovine Somatotropin (BST), also called bovine growth hormone (BGH), has been used to increase milk production in dairy cattle by 10 to 15 percent (BST is also naturally produced by the cow's pituitary gland). BST is not typically found in the waste-stream of animals fed BST (Harter 2006, personal communication).

Some hormones are known or suspected of interfering with the endocrine system of various species, including humans, and can affect the reproductive biology of fish and other aquatic vertebrates species. Although steroids and other hormones adsorb to soil and sediment, they are often sufficiently mobile to impact water quality and have been reported in surface and groundwater following cattle manure and poultry litter land application (Shore, Correll, and Chakraborty 1995).

Fish farming is an increasing source of drugs in the environment. In aquaculture, the drugs are applied as feed additives and discharged directly into water (Baticados and Paclibare 1992, Migliore et al. 1993). Through overfeeding, loss of appetite by diseased fish, and poor adsorption of the drugs, an estimated 70 percent of the drugs administered are released into the environment (Schneider 1994). Those substances used to fight bacterial diseases, when released to the environment, can potentially promote development of bacterial resistance in the general microbial population in the environment.

Salts

Significant amounts of sodium and potassium may be present in dissolved mineral salts that pass undigested through animals after consumption of feed. Calcium, magnesium, chloride, sulfate, bicarbonate, carbonate, and nitrate may also contribute to the salinity of manure. Bicarbonate and carbonate are also produced by mineralization of organic waste.

Relatively low levels of salts can increase blood pressure in sensitive individuals, thus increasing their risk of heart attacks and strokes. Few drinking water treatment processes are capable of removing salts. Increasing salinity can disrupt the balance of fresh water ecosystems, making it difficult for native species to thrive. Dairy cattle wastes have been blamed as the cause of elevated salinity levels in groundwater (Harter, Meyer, and Mathews 2002; Harter et al. 2002). Wastes with high salt levels that are applied to land can also contaminate groundwater, adversely impact soil structure, reduce soil permeability, and reduce crop yields (USEPA 2003a).

Solids

Solids from animal wastes include manure itself and other components that have been mixed with it. In terms of impacts on public water supplies, solids provide a medium for the accumulation and transport in surface waters of other pollutants such as pathogens, nutrients and trace elements. They may also adversely affect the taste and odor of source waters. Increased solids loadings may at times prove difficult to remove for some water treatment facilities. Solids also can cause environmental effects by increasing the turbidity of surface waters, thus limiting the growth of aquatic plants that serve as habitat for fish, shellfish, and other aquatic organisms. Settled solids can interfere with the habitat of bottom-dwelling organisms (USEPA 2003a).

Organic Matter

Animal manure contains many biodegradable organic (carbon-based) compounds. These compounds, once in surface water, tend to be decomposed by bacteria and other microorganisms, with a resulting decrease in aqueous dissolved oxygen concentrations. This in turn can have deleterious effects on aquatic organisms that require oxygen to survive, and severe reductions in surface water dissolved oxygen can result in hypoxic zones and even fish kills. Less severe drops in dissolved oxygen levels can result in decreased biodiversity and a relative increase in species that require lower levels of oxygen (USEPA 2003a). The degraded organic matter can also potentially react with drinking water disinfectants (e.g., chlorine) to produce potentially harmful disinfection byproducts.

Volatile and Odorous Compounds

Volatilization of manure compounds and their degradation products (e.g., ammonia, hydrogen sulfide, methane) can result in subsequent redeposition on surface waters and land. Deposition of atmospheric ammonia can potentially contribute to eutrophication of surface waters. Odors from AFOs are a common complaint from neighbors.

Heavy Metals and Trace Elements

Among the trace metals in animal waste of environmental concern are arsenic, copper, selenium, zinc, cadmium, molybdenum, nickel, lead, iron, manganese, aluminum, and boron. In particular, arsenic, copper, selenium, and zinc are often used in poultry, swine and bovine diets as growth stimulants or to supplant the use of antibiotics. As an example, arsenic-containing feed supplements are added to poultry diets. Pesticides used directly on livestock to control flies and other pests can also be a contributing source of trace metals. Nearly all of these additions end up in manure. Zinc and copper solutions are often used as a foot bath for prevention of hoof rot in dairy cattle, and thus can greatly increase the concentrations of these metals in dairy cattle manure. USEPA (2003a) does not consider trace elements in agriculturally-applied manure to pose much of a risk to human health and the environment, though repeated application of manure above recommended rates could potentially result in cumulative metal loadings that could have potential effects on human health and the environment.

Trace metals tend to persist in the environment and can bioaccumulate in animal and plant tissues. Copper and zinc in some manures even exceed levels in biosolids. More consideration could be given to the potential increase in management costs from metals in manures compared to savings from supplementing diets with copper, zinc, or arsenic levels above nutritional requirements (Chaney 2003). Arsenic in poultry litter is readily soluble, and this solubility can be enhanced by a transformation of arsenic to more soluble arsenate during composting (Garbarino et al. 2003).

Pesticides

Pesticides are sometimes used directly on livestock to control houseflies and other pests. Some pesticides can be linked with endocrine disruption effects in a variety of species. USEPA (2003a) notes that little research has been reported on the runoff of pesticides from land where manure has been applied, though they mention one study that showed such runoff (the pesticide cyromazine used to control flies in poultry litter) can increase with the rate of manure and litter applied and the intensity of subsequent rainfall events (USEPA 2003a).

CHAPTER 5
STRATEGIC CONTROL MEASURES FOR AFOs

Substandard operational practices at animal feeding operations can result in routine, low-level chronic releases of contaminants to the environment, and even well-run "controlled" processes can still produce routine releases. In addition, there may be accidental acute releases of contaminants, especially when exacerbated by other factors such as extensive rainfall events. Even well-run operations are not immune to extensive rainfall and flooding. Accidental, acute releases can often be large, but are typically infrequent. In the long run, the chronic daily releases, especially from uncontrolled and poorly operated facilities, may release as much or more contaminants of concern to water quality.

Farms should have, as appropriate, a Whole Farm Plan (WFP) that discusses all farm operations including waste disposal, a Comprehensive Nutrient Management Plan (CNMP) that focuses on manure and nutrient management, and an emergency action plan that provides contingency planning for foreseeable emergencies. Planning should include a pollutant mass balance analysis to assess if the farm is a significant contributor of pollutants to the environment. Emergency action plans should detail contingencies for potential major events such as large-scale spills of liquid manure, lagoon water, fuel oils, and other liquid materials.

The production of unnecessary wastes should be kept to a minimum, and management procedures for the waste produced should be based on both environmental and economic considerations. A wide variety of Best Management Practices (BMPs) can be used to reduce or control animal waste contaminant releases from AFOs and CAFOs. Proactive solutions focus on reducing the amount of contaminant waste created (e.g., feed type optimization, run-on control, animal watering practices, fencing to keep animals out of streams, etc.) and reactive solutions for managing and treating wastes (e.g., runoff control, collection, storage and treatment of wastes, etc.). Solutions include both "natural" (e.g., vegetative buffer zones) and man-made solutions (e.g., waste treatment). Approaches can include physical BMPs, nutrient management plans for applying animal waste to croplands and pasture, and stakeholder education and outreach. Approaches will depend largely on the type and number of animals, the local hydrogeology and climate, and the applicable regulations. Assessing the effectiveness of BMPs can at times be difficult because in many cases the physical reality is extremely complex, and in addition funding is often limited for such assessments.

This Chapter identifies and characterizes the major types of BMPs applicable for AFOs and other related agricultural operations to reduce, eliminate, or control contaminants of concern to source water quality, and provides the reader with reference sources for accessing more detailed information. These are identified and characterized so that water utility operators can be aware of these solutions and encourage their use in discussions with AFOs and CAFOs, as well as their regulators.

A summary of select management practices for animal confinement areas are listed in Table 5.1, and for cropland in Table 5.2. Several case studies that use many of these BMPs are described in Chapter 7. Descriptions of typical waste production, management and treatment strategies are provided in Chapter 4 for each of the different major CAFO types (beef cattle, dairy cattle, swine, poultry, and aquaculture).

Table 5.1
Example management practices for animal confinement areas
(Derived from Table 1 in Telega and Cook 2003; www.lpes.org/cafo/05FS_AFO.pdf)

Environmental goal	Management practices
Animal nutrition management	• Design feed diets to not exceed nutrient requirements of the animals, thus reducing the amount of excreted nutrients (e.g., low-phosphorus feeds) • Food supplements can be used to reduce production of specific waste constituents (e.g., zinc in poultry diets can reduce ammonia emissions (Kim and Patterson 2003)) • *For more info, see LPES Lessons #10 to 13 on animal dietary strategies (www.lpes.org/les_plans.html)*
Clean water diversion - exclude clean water from contaminated areas	• Runoff water from roofs, driveways, and other clean areas should not be allowed to mix with water contaminated with manure or feed. • Surface runoff from areas upgradient of the animals should be channeled around the facility • Keep clean water clean; collect and treat contaminated water • *See LPES Lesson #22 on Open Lot Runoff Management Options*
Eliminate, contain, collect, and treat runoff	• Use covered facilities for storing waste, feed, or fertilizer • Re-evaluate the use of the barnyard. Eliminating the barnyard, making it smaller, or relocating it to a better site will make addressing runoff concerns less problematic and expensive. Replace barnyards with covered concrete pads. • Barnyard and other facility runoff should be contained, collected, and released to vegetative filter areas or spread on cropland • *See LPES Lesson #22 on Open Lot Runoff Management Options*
Control runoff from bunk silos	• Consider installing a low-flow, high concentration collection system with a high-flow, low-concentration filter area one-third the size of the bunk area to control silage juices • Managing the silage for the proper moisture content and to prevent spoilage also reduces the potential of pollution
Identify alternative uses for manure	• Haul manure to areas outside of the source water area where it is needed and won't help cause excessive nutrient levels in source waters • Consider value-added salable products that can be generated from animal manure, including composted and odor-free fertilizers, soil amendments or feed additives • Energy can be obtained from manure through either combustion, biological conversion (anaerobic digestion), or chemical conversion (to methanol) • See www.epa.gov/npdes/pubs/cafo_report.pdf

(continued)

Table 5.1 (Continued)

Environmental goal	Management practices
Waste treatment	• Lagoons, anaerobic digesters, composting, physicochemical process systems, and constructed wetlands are just a few of the available waste treatment options • Apply good operating and maintenance procedures to treatment systems such as for milk house waste or egg wash water • Pump all settling and storage tanks regularly • Keep leach fields and vegetative filter areas clean, healthy, and functioning
Fence animals out of watercourses	• Do NOT allow animals to wander along creeks and streams! • Install fencing and provide alternate drinking water supply for the animals • Use cattle crossings where needed • Maintain grass buffer areas around lakes and ponds, and along creeks that run close to the farm
Keep piles of manure and spoiled silage away from watercourses	• Allow at least 300-foot flow path to the nearest down-slope watercourse • Manage the flow course to provide diffuse overland flow through well-vegetated fields • Keep upslope water from contacting the piles
Restrict access of animals to well heads	• Wells are in direct contact with ground water and can easily cause contamination when animals mill around them, particularly old, shallow, or abandoned wells
Properly dispose of dead animals	• Consult authorities and comply with state and local laws about mortality disposal • Arrange for rendering service pickup within 48 hours • Proper composting of dead animals is the next best option • If buried on farm, keep at least 200 foot away from a watercourse • Bury under a 3-foot minimum of well-packed earth • *See LPES Lesson #51 on Mortality Management*
Odor and nuisance control	• Unpleasant odors are the number one reason neighbors complain about AFOs. AFOs should do everything operationally possible to keep manure odors and other nuisances to a minimum • Maintain open communication with neighbors so complaints come directly to you instead of local authorities • *See LPES Lessons #40 to #44 on Outdoor Air Quality*

Note: All Livestock and Poultry Environmental Stewardship (LPES) Curriculum lessons are available at www.lpes.org/les_plans.html

Table 5.2
Example management practices for cropland
(Derived from Table 2 in Telega and Cook 2003; www.lpes.org/cafo/05FS_AFO.pdf)

Environmental goal	Management practices
Reduce soil erosion	• Maintain riparian buffers alongside watercourses and edges of field and barnyard areas • Apply appropriate crop rotations to reduce erosion • Irrigation water should be managed to optimize the amount of water applied and maximize the efficiency of where and when it is applied in relation to crop needs • Apply appropriate water management practices on fields with persistent gully erosion • Conservation tillage - plant residues from previous crop cycles are not completely plowed into the soil, and instead remain on top of the soil to provide cover and reduce runoff (USEPA 2001a, USEPA 2001b). This practice can be especially helpful for reducing phosphorus losses
Maintain field water management practices	• Irrigation water should be managed to optimize the amount of water applied and maximize the efficiency of where and when it is applied in relation to crop needs • Ensure earthen diversions and grass waterways are properly vegetated, are free of sediment buildup, and do not show signs of water channeling
Know the phosphorus and nitrogen status of the fields	• Soil test regularly • Reduce manure and fertilizer phosphorus on High or Very High testing fields • Over two-thirds of the phosphorus entering the farm as feed ends up in the manure. Keep feed ration phosphorus at recommended levels. • For non-ruminants, feeding low phosphorus or the enzyme phytase may result in less total phosphorus in feed rations. • *For more info, see the LPES Lessons 10, 11, 12, and 13 on Animal Dietary Strategies and Lesson #34 on Phosphorus Management for Agriculture and the Environment (www.lpes.org/les_plans.html)*
Use pre-sidedress nitrogen tests (PSNT). PSNT is used for optimizing N fertilizer use, and is based upon the timely measurement of soil nitrate in the top foot of soil just before the corn crop starts its rapid period of nitrogen uptake at the 4 to 6 leaf stage of corn (6 to 12 inches tall)	• Employ the test on fields receiving manure that have been in corn for 2 years or more and where pre-plant broadcast nitrogen is not used • The test determines the need for additional nitrogen at sidedress time • Particularly useful on fields that have previously received organic sources of nitrogen such as manure or biosolids. *For more info, see http://nmsp.css.cornell.edu /publications/ PSNT2005.pdf* or *www.agnr.umd.edu/usersagron/nutrient/ Resources/ PSNT.htm*
Plant cover crops on soybean and corn for silage fields	• Cover crops reduce soil erosion by reducing the effects of raindrop impact on open ground • Cover crops add to soil organic matter and capture some nitrogen from fall applications of manure, reducing leaching of the winter thaw

(continued)

Table 5.2 (Continued)

Environmental goal	Management practices
In Northern climates, kill sods later in the fall	• While soils are still warm in early fall (above 40°), sod residues, especially legumes, will start decomposing, releasing inorganic nitrogen. If little active uptake occurs from the killed sod, the released nitrates will leach during fall and early winter rains • Cold soils of late fall retard decomposition and keep nitrogen in the organic, non-soluble form
Know the nutrient content of the manure	• Test manure annually to be confident about its nutrient value for crop production • Maintain a record of past manure analysis • *See LPES Lesson #31 on Manure Utilization Plans*
Calibrate manure application equipment regularly	• Knowing the rate of manure application to land is critical information for effectively utilizing manure as a crop nutrient and reducing fertilizer inputs for crop production • *See LPES Lesson #32 on Land Application Best Management Practices*
Practice uniform manure spreading	• Understand the limitations of the equipment to apply uniform layer of manure • Emphasize the importance of evenness and uniformity to all workers applying manure to your fields • *See LPES Lesson #32 on Land Application Best Management Practices*
Determine fields that have a high potential for runoff	• Avoid spreading on these fields during wetter times of the year • Use a Phosphorus Index or other risk assessment tools to understand the risks of manure applications during wet and other sensitive seasons • *See LPES Lesson #33 on Selecting Land Application Sites*
Spread manure away from wells, springs, and watercourses	• Keep manure at least 100 feet from wells and springs to reduce the potential contamination of recharge areas. • Maintain vegetative buffers along waterbodies in fields receiving manure • Contact your County Cooperative Extension, Department of Health, or County Soil and Water Conservation District to learn about any local restrictions to land application of manure • *See LPES Lesson #33 on Selecting Land Application Sites*
When possible, till in fall-applied manure	• Such incorporation will break up soil macropores and reduce preferential flow into drain tile or shallow groundwater • Incorporation reduces the potential of manure leaving the field through overland flow

Note: All Livestock and Poultry Environmental Stewardship (LPES) Curriculum lessons are available at www.lpes.org/les_plans.html

A vast amount of information on agricultural and animal waste management planning and best management practices is available. The list of references and web links provided below in Table 5.3 can guide the reader to some of the more reliable sources of that available information that are available on the internet. The project website (www.eetinc.com/cafo) was developed to readily lead the user to this information.

Table 5.3
Internet links for information on agricultural and animal waste BMPs

Title	Author and description	Website
USEPA National Agriculture Compliance Assistance Center	USEPA website - information on animal production practices and BMPs AFO Best Management Practices	www.epa.gov/agriculture www.epa.gov/agriculture/anafobmp.html
Best Management Practices and Controls	USEPA - AFO Virtual Information Center. Links to web site and documents highlighting best management practices related to controlling discharges from AFOs	http://cfpub.epa.gov/npdes/afo/afoinfo.cfm?view=category&link_cat=27
Nutrient Management	USEPA - AFO Virtual Information Center. Information and tools to help develop nutrient management plans, including state technical standards	http://cfpub.epa.gov/npdes/afo/afoinfo.cfm?view=category&link_cat=25
Managing Manure Nutrients at Concentrated Animal Feeding Operations	Technical guidance designed to supplement the NPDES Permit Writers' Guidance Manual and Example NPDES Permit CAFOs. Provides information on complying with permit requirements, and technologies and management practices that may both improve the production efficiency of CAFOs and further protect the source water quality. (EPA-821-B-04-009; 12/01/2004).	www.epa.gov/npdes/pubs/cafo_manure_guidance.pdf
National Management Measures for the Control of Nonpoint Pollution from Agriculture	USEPA (2003d) - technical guidance and reference document containing information on the best available and economical achievable means of reducing pollution of surface and groundwater from agriculture	www.epa.gov/nps/agmm/
CNMP Watch	Guidance for developing CNMPs (website funded by USEPA)	www.cnmpwatch.com/
Risk Management Evaluation For Concentrated Animal Feeding Operations	USEPA (2004b). Provides a good recent review of the impact of beef cattle, dairy, swine and poultry operations on the environment	www.epa.gov/ORD/NRMRL/pubs/600r04042/600r04042.pdf

(continued)

Table 5.3 (Continued)

Title	Author and description	Website
Managing Livestock, Poultry, and Horse Waste to Prevent Contamination of Drinking Water	USEPA (2001a). Source Water Protection Practices Bulletin. Fact sheet highlighting some effective source water contamination prevention measures related to livestock, poultry, and horses	www.epa.gov/safewater/protect/pdfs/livestock.pdf
Managing Agricultural Fertilizer Application to Prevent Contamination of Drinking Water	USEPA (2001b). Source Water Protection Practices Bulletin. USEPA. 2001. EPA-916-F-01-028. This fact sheet addresses some source water contamination prevention measures related to agricultural fertilizer application	www.epa.gov/safewater/protect/pdfs/fertilizer.pdf
Alternative Technologies/Uses for Manure (Draft Report)	USEPA (2002b). A compilation of alternative uses for manure from AFOs, including conversion of manure to value-added products (e.g., as fertilizers, soil amendments or feed additives) and as an energy source.	www.epa.gov/npdes/pubs/cafo_report.pdf
Agricultural BMP Effectiveness Database	PSU (1992). Developed for USEPA by Pennsylvania State University and others.	http://iaspub.epa.gov/edr/edr_proc_qry.navigate?P_LIST_OPTION_CD=CSDIS&P_REG_AUTH_IDENTIFIER=1&P_DATA_IDENTIFIER=90637&P_VERSION=1
BMP Database for Nutrient Reduction	Yagow et al. (2002). Developed by Virginia Polytechnic Institute and State University for the USEPA Chesapeake Bay Program.	http://asae.frymulti.com/abstract.asp?aid=7565&t=2
International Stormwater BMP Database	ASCE/USEPA-supported database includes numerous reports of stormwater BMP efficiency measurements.	www.bmpdatabase.org/
Nutrient and Pest Management	USDA-NRCS website with nutrient management tools and information	www.nrcs.usda.gov/technical/nutrient.html
Agricultural Waste Management Field Handbook	USDA (1999). Large NRCS guidance document covering many relevant subjects in detail, including agricultural waste management systems for each major type of AFO. See Chapters 9 and 10 in particular.	www.wcc.nrcs.usda.gov/awm/awmfh.html
Animal Waste Management	USDA-NRCS website - Nutrient management tools and information, including software for animal waste management planning	www.wcc.nrcs.usda.gov/awm/

(continued)

Table 5.3 (Continued)

Title	Author and description	Website
NRCS National Handbook of Conservation Practices	USDA-NRCS. List of over 150 National Conservation Practice Standards	www.nrcs.usda.gov/technical/Standards/nhcp.html
Buffer Strips: Common Sense Conservation	USDA-NRCS website with information on buffer strips	www.nrcs.usda.gov/feature/buffers/
Draft CNMP Technical Guidance	USDA-NRCS National Planning Procedures Handbook (Subpart E, Parts 600.50-600.54 and Subpart F, Part 600.75). Handbook with guidance for Comprehensive Nutrient Management Planning	www.nrcs.usda.gov/programs/afo/cnmp_guide_index.html
LPES Curriculum Lessons	Livestock and Poultry Environmental Stewardship (LPES) Curriculum guidance for agricultural environmental stewardship	www.lpes.org/les_plans.html
CAFO Fact Sheets	Series of excellent fact sheets on CAFO Rule requirements, NMPs, and BMPs	www.lpes.org/CAFO_PDF_Land.html
Cornell Manure Management Program	Cornell University website with fact sheets, case studies, and technical information for dairy manure management and treatment systems	www.manuremanagement.cornell.edu
Managing Dairy Manure in the Central Valley of California	University of California report on managing dairy manure	www.waterboards.ca.gov/centralvalley/available_documents/dairies/uc-committee-of-experts-rpt.pdf
Manure Matters Web Site	University of Nebraska Cooperative Extension - includes workbooks on manure application, storage, and odor management	www.ianr.unl.edu/manure
Animal Operations	North Carolina State University and North Carolina A&M State University – links to various Biological and Agricultural Engineering Extension publications and guidance	www.bae.ncsu.edu/programs/extension/publicat/wqwm/animops.html
Ohio Livestock Manure and Wastewater Management Guide	Ohio State University Extension (OSU bulletin 604) – information about manure and wastewater management and treatment	http://ohioline.osu.edu/b604/

(continued)

Table 5.3 (Continued)

Title	Author and description	Website
Purdue Animal Manure Solutions	Purdue University site that has lots of web links and articles on manure management	http://danpatch.ecn.purdue.edu/~epados/pams/
Best Management Practices for Phosphorus in the Environment	University of Tennessee Agricultural Extension Ser vice – good description of phosphorus issues and BMPs	www.utextension.utk.edu/publications/pbfiles/PB1645%20.pdf
Managing Manure Phosphorus to Protect Water Quality	University of Missouri Extension (John A. Lory) - information on managing manure phosphorus	http://muextension.missouri.edu/explore/agguides/soils/g09182.htm
Managing Phosphorus from Animal Manure	Oklahoma Cooperative Extension Service – information on managing manure phosphorus	http://osuextra.com/pdfs/F-2249web.pdf
Best Management Practices for Phosphorus	Kansas State University Agricultural Experiment Station and Cooperative Extension Service	www.oznet.ksu.edu/library/crpsl2/mf2321.pdf
Dairy/AFO Environmental Quality	University of California at Davis, Groundwater Cooperative Extension Program – list of websites related to animal waste management	http://groundwater.ucdavis.edu/dairy_links.htm
Dairy Nutrient Management and Groundwater Quality	University of California at Davis, Groundwater Cooperative Extension Program – description of research project looking at dairy operation impacts on groundwater quality	http://groundwater.ucdavis.edu/gw_201.htm
Farm*A*Syst	National Farmstead Assessment System (supported by USDA CSREES, USDA-NRCS, and USEPA) – fact sheets and worksheets for agricultural waste management	www.uwex.edu/farmasyst/index.html
Conservation for Agriculture's future	Conservation Technology Information Center (Purdue University) website with brochures and links to information on agricultural BMPs	www.ctic.purdue.edu/Core4/Core4Main.html
Midwest Plan Service	Iowa State University listing of publications for agriculture	www.mwpshq.org/
BMPs: Land Application of Animal Manure	Ohio State University Extension (10 page summary)	http://ohioline.osu.edu/agf-fact/0208.html

(continued)

Table 5.3 (Continued)

Title	Author and description	Website
Liquid Manure Application Methods	Colorado State University Cooperative Extension (5 page summary)	www.ext.colostate.edu/pubs/livestk/01223.html
Idaho Waste Management Guidelines for Aquacultural Operations	Waste management systems for aquaculture facilities are discussed in detail by Idaho DEQ (2002)	www.deq.idaho.gov/water/prog_issues/waste_water/pollutant_trading/aquaculture_guidelines.pdf
Southern Regional Aquaculture Center	The SRAC website provides a series of fact sheets about aquaculture and aquaculture waste management	www.msstate.edu/dept/srac/
Waste Management in Aquaculture	Miller and Semmens (2002) - summary of aquaculture facilities, including BMPs to reduce aquaculture waste	www.wvu.edu/~agexten/aquaculture/waste02.pdf
Effects of Voluntary Agricultural Best Management Practice Implementation on Water Quality	This literature review from researchers at The University of Georgia summarizes 21 studies that supplied some indication of the effectiveness of voluntary programs and a summary to provide some interpretations of the results	http://www.engr.uga.edu/service/extension/agp2/resources/publication/A-G/Effects%20of%20Voluntary%20A18.doc
Source Water Protection Reference Manual	CDM (2002) - a CD-rom reference tool for water utilities with information on four main facets of watershed protection: BMPs, planning, regulations, and water supplier experience. Easy to use search engine	www.awwarf.org/research/TopicsAndProjects/projectSnapshot.aspx?pn=2651

As examples, typical animal waste management systems for each major type of AFO are described in detail in Chapter 9 of the USDA Agricultural Waste Management Field Handbook (USDA 1999) and in the Ohio Livestock Manure and Wastewater Management Guide (http://ohioline.osu.edu/b604/). For descriptions of design criteria for many other possible animal waste handling and treatment methodologies see Chapter 10 of the USDA Agricultural Waste Management Field Handbook (USDA 1999).

Pyke et al. (2003) performed a literature review of agricultural BMP efficiency data, and conducted a modeling investigation assessing the impacts of major point and non-point sources on raw water treatability. Other compilations of BMP efficiency data may be found in the BMP Database for Nutrient Reduction (Yagow et al. 2002) developed by Virginia Polytechnic Institute and State University for the USEPA Chesapeake Bay Program, and the Agricultural BMP Effectiveness Database (PSU 1992) developed for USEPA by Pennsylvania State University and others. The ASCE/USEPA International Stormwater BMP Database includes numerous reports of stormwater BMP efficiency measurements (www.bmpdatabase.org). A summary of the fate of *Cryptosporidium* and the effectiveness for *Cryptosporidium* inactivation via different BMPs (e.g., composting, anaerobic digestion, manure slurry storage) was provided by Finstein (2004).

Water utilities always need to bear in mind that any efforts expended by the AFO facility to reduce potential impacts of wastes on water supply will require expenditures of time, money, effort, and training; consequently, these will be viewed by the AFO operator as costing them potential profits. Economic trends in the industry will impact the ability of operators to pay for environmental improvements, and few farms have sufficient money to build treatment systems. Targeting effective BMPs to the highest priority pollutant source areas is needed to maximize the impact and value of funds expended for source water protection.

Costs for agricultural BMPs are provided by the USEPA (2003d), USDA (www.nrcs.usda.gov/technical/RESS/issues/bufferecon.html), and Kansas State University (2003). Potential funding sources for projects related to source water protection and agricultural practices are identified and discussed in Chapter 6.

Effort was made to identify and describe below alternatives that do not necessarily adversely impact the economics of the animal feeding operations. For example, alternatives are identified that utilities can point out to AFOs and their regulators that do not negatively impact or cost more than current AFO operations, but can significantly reduce contaminant releases. In other cases, there are strategies that AFOs can employ that end up reducing contaminant emissions while also positively impacting AFO operations and profits. Cost-sharing programs have proven effective for helping motivate farmers to make pollution control BMP improvements.

WASTE MANAGEMENT PLANNING

The USEPA CAFO Rule requires a site-specific Comprehensive Nutrient Management Plan (CNMP) be prepared for subject CAFOs, including a plan for land application if that is done. The USEPA and USDA recommend that all AFOs develop and implement a site-specific, economically feasible, and technically sound CNMP. AFOs not defined as CAFOs are encouraged to participate in voluntary programs conducted by federal, state, and local agencies to learn about financial and technical assistance available for developing and implementing a CNMP for their operation. In many cases the nutrient management plans and manure land application schedules are completed with assistance from the USDA-NRCS, Soil and Water Conservation Districts, and other similar organizations.

All farms and AFOs should have management plans that consider the characteristics of all the different wastes and byproducts generated, means to reduce the amount of waste produced, measures to adequately protect the environment (including drinking water sources) from contamination from the wastes, and economic and safety considerations. A complete analysis would include the type, quantity, consistency, location, and timing of the waste produced (USDA 1999), and include measures to minimize nutrient excretion from the animals, ensure proper storage of manure and manure-contaminated water, specify the land application of manure nutrients at agronomic rates that do not adversely affect source water quality, and require maintenance of appropriate records. Seasonal variations in the amount and rate of waste production, as well as in the available storage, treatment and disposal options, should also be considered. Farm employees should be trained in the latest procedures for all plans. Confidentiality of the components of CNMPs can be an issue, as many producers won't want their plans to be available to the public through the Freedom of Information Act.

PROACTIVE SOLUTIONS FOR REDUCING THE AMOUNT OF CONTAMINANT WASTE CREATED

For most any type of activity, it is better and more economical to reduce the amount of waste created, or take predetermined measures to favorably alter the type or characteristic of waste produced. Animal feeding operations are no exception to this rule. Select common examples of this type of approach are discussed below.

Feed Type Optimization

Feed characteristics and quantity should be monitored to optimize the needs of the animals and minimize costs to the producer. Nutrients such as phosphorus and nitrogen are essential for animal growth. Unfortunately, most animal wastes, including chicken litter, have an abundance of phosphorus versus nitrogen when compared to plant requirements. One strategy is to match the feed formulation to the age (production stage) of the animals to match nutrient requirements as closely as possible (and thus minimize the amount excreted). Diets without excess phosphorus can help to make the N/P balance in the waste closer to the optimum balance for use as a fertilizer, and thus reduce phosphorus loadings to fields. The addition of phytase to chicken and hog feed helps make phosphorus more bioavailable to the animals, thus allowing lower amounts of phosphorus to be used in the feed.

As another example, chemical treatment of waste (Nahm 2003) or use of feed supplements containing zinc (Kim and Patterson 2003) has reduced ammonia emissions during poultry operations. This has not only reduced the potential for release of this possible contaminant, but has also improved poultry production (not surprisingly, poultry grow more productively when not in presence of ammonia).

The Livestock and Poultry Environmental Stewardship (LPES) Curriculum provides a series of articles on animal dietary strategies at www.lpes.org/les_plans.html. Another good source of information on dietary strategies is Chapter 8 in the USEPA Development Document for the Proposed Revisions to the NPDES Regulation and the Effluent Guidelines for CAFOs (USEPA 2002c).

Clean Water Diversion

Clean water diversion and run-on control are effective measures for preventing precipitation or surface runoff in upgradient areas from reaching an animal waste handling site (USEPA 2001a). In other words, the goal is to keep clean water clean. Stormwater management both inside and outside feedlots and livestock yards is essential to prevent groundwater and surface water contamination. Facilities should use physical means to prevent stormwater from unnecessarily contacting areas where animals are present or where manure or feed is stored. Reducing the size of open holding areas, using covered facilities for housing animals and storing waste, and directing uncontaminated stormwater runoff away from these areas all help at minimizing pollution.

Runoff from upgradient of the facility should be directed around the areas with manure. Surface diversions, such as trenches, earthen ridges or diversion terraces built above the feedlot or barnyard, can be used to direct surface flow from upgradient areas away from areas where animal waste is located. Runoff water from roofs, adjacent lands, driveways, and other clean areas should not be allowed to contact manure or feed or mix with water contaminated with manure or feed. Rain gutters and downspouts on animal shelter roofs keep runoff clean by controlling and diverting precipitation away from animal waste. Proper protection (or isolation) of agricultural drainage well inlets is important for guarding against ground water contamination (drainage well inlets are injection wells used to drain excess agricultural irrigation water from farm fields). For more information, see LPES Lesson #22 on 'Open Lot Runoff Management Options' (www.lpes.org/Lessons/Lesson22/22_Lot_Runoff.html).

Animal Watering

Livestock housed in open areas (e.g., pastures, and open lots and barnyards) should not have direct access to a creek or stream where they can defecate directly into the waterbody and cause physical damage to stream banks and bank vegetation. Livestock fencing and stream cattle crossings should be used to keep pasture animals out of surface waters, and mechanical livestock watering systems can be used as replacement drinking water sources for the animals. These are popular agricultural BMPs used in many source water protection programs. Fence types include standard or conventional (barbed or smooth wire), suspension, woven wire, and electric fences (USEPA 2001a).

The USDA conducted a study of four different types of novel livestock watering alternatives. The Elkhorn Creek BMPs Demonstration Project assessed ram pumps, pasture pumps (cattle activated pump), a solar-powered water pump, and limited access watering points (using modern electric fencing components). For more details on this study see www.epa.gov/region04/water/nps/projects/ky93-1b.htm.

Many swine and poultry AFOs use nipple-type watering systems so that the animals spill as little water as possible. This reduces the amount and moisture content of the manure or litter waste from those shelters.

Animal Health Management

Sick animals are a greater source of microbial pathogens than healthy ones. Proper health care measures not only help preserve the health of the animals and prevent the spread of disease, but also reduce the potential health risk from the manure. For example, since calves are a primary source of *Cryptosporidium* oocysts and *Giardia* cysts, good calf health is important for minimizing the presence of the pathogen.

Soil Erosion Prevention

Soil erosion practices have been well studied for years, and are well-known to the agricultural community. Effective soil conservation practices not only preserve the soil resource for agricultural use, but reduce the amount of sediment and associated pollutants such as sorbed organic material and attached pathogenic microorganisms and phosphorus transported to surface waters.

Appropriate tilling, planting, and watering practices should be used to minimize the amount of soil particles transported off site. Crop rotations and planting cover crops can also help. Riparian buffers between plowed fields and surface waters and drainage channels can be very effective at reducing sediment loads to the waters. Earthen diversions and grass waterways should be properly vegetated, free of sediment build up, and have no signs of water flow channeling (Telega and Cook 2003). Vegetated filter strips for reducing erosion are discussed in more detail below. Soil erosion best management practices are discussed in numerous sources, including the following:

- USDA Agronomy and Erosion Technical Resources: www.nrcs.usda.gov/technical/agronomy.html
- NRCS National Handbook of Conservation Practices: www.nrcs.usda.gov/technical/Standards/nhcp.html
- USEPA - On-line graphic education module about soil erosion processes and BMPs: www.epa.gov/seahome/erosion.html
- Soil Management and Conservation Practices - Iowa State University Extension website: http://extension.agron.iastate.edu/soilmgmt/

Conservation Tillage and Crop Rotation

Conservation tillage is a technique where crops are grown with minimal cultivation of the soil. With less tillage than traditional planting methods, plant residues from previous crop cycles are not completely plowed into the soil, and instead remain on top of the soil to provide cover and reduce runoff (USEPA 2001a, USEPA 2001b). This practice can be especially helpful for reducing phosphorus losses (USEPA 2001b). However, with no tilling, worm holes can provide preferential flowpaths for manure to reach underlying subsurface drainage systems of groundwater. Subsurface tiles are frequently used for subsurface drainage in fields in the Midwestern US east of the Mississippi River, and to a lesser degree in some eastern states.

Crop rotation can help minimize fertilizer and pesticide needs, and thus not only reduce potential transport of these chemicals to environments outside of the farm, but also provide crop improvement and economic benefits. Different plants have different nutrient uptake requirements and abilities, and occasional switching of plant type helps to reduce the amount of fertilizer that must be applied. Legumes can be planted to provide nitrogen for subsequent crops. Switching to deep-rooted crops can help provide uptake of nutrients deeper in the soil where shallow-rooted plants can't reach (USEPA 2001b). Information on tillage practices is available from the USEPA at www.epa.gov/agriculture/ag101/cropsoil.html#envconcerns.

Fertilizer and Feed Handling

Efforts should be made to handle feed and fertilizer stocks so that production of unnecessary waste is kept to a minimum. All such material should be stored indoors or in covered areas to avoid contact with precipitation. Reducing the amount of spilled feed can help reduce potential waste generation as well as save the farm operator money. Good management and maintenance of feeders, watering facilities, feed storage facilities, and associated equipment will help reduce feed waste (USDA 1999).

Fertilizer type and rates of application should be designed to meet, but not substantially exceed, plant requirements. Fertilizer and feed storage and mixing facilities should be protected from spills, leaks and infiltration by precipitation. Impermeable floors for storage facilities help to reduce product loss and soil contamination, and secondary containment structures may be used for liquid product storage areas. Handling of fertilizer and feeds should be conducted away from wellheads and surface waters (USEPA 2001b).

REACTIVE SOLUTIONS FOR MANAGING AND TREATING WASTES

Once a waste is generated, it must be collected, transported, treated, recycled, or disposed of. Various solutions are available, and site-specific factors need to be taken into consideration in selecting and designing these systems, including the type, amount and characteristics of the waste, regional climatic conditions, available treatment and disposal options, labor requirements, and costs. Various BMPs for animal waste management applicable to AFOs and CAFOs are discussed below.

Runoff Control

A large part of the waste associated with many livestock operations includes contaminated runoff from open animal holding areas. Reducing the amount of precipitation that contacts animal waste is a first step, and managing the stormwater that does contact manure is also essential. As discussed above, runoff from upgradient of the farm facility should be directed around the areas with manure. Livestock housing and related buildings should have gutters to keep rainwater out of the manure waste stream. Runoff from holding areas can be directed into vegetative buffer strips, grass buffers, or directed overland or in perforated pipes underground to allow soil treatment of the runoff. Different methods are used to spread the runoff evenly into the filter areas; these can include gravel trenches and trenches filled with bark or wood chips. These trenches can also provide some treatment benefit, such as microbial degradation or organics by biofilm on the gravel, or sorption of contaminants onto bark or wood. In some cases, such as outdoor cattle feed lots, runoff from the holding pens is directed downhill to a storage lagoon.

Runoff control is well described in Chapter 10 of the USDA Agricultural Waste Management Field Handbook (USDA 1999), and in various fact sheets provided by the LPES Curriculum (CAFO Production Area Fact Sheets at www.lpes.org/CAFO_PDF_Prod.html; and Lesson 22 on Open Lot Runoff Management Options, www.lpes.org/Lessons/Lesson22/22_Lot_Runoff.html).

Waste Collection and Transfer

Farm management plans should include descriptions of appropriate waste collection and transfer techniques designed to minimize the release of contaminants to the environment. This includes the initial gathering of the waste from the point of origin to a collection point, and subsequent transfer for further treatment or disposal. The waste may be transferred as a solid, liquid, or slurry, depending on its solids concentration. These plans should identify the consistency of the waste, the collection and transportation methods, collection point locations, scheduling of waste collection, labor requirements, required equipment or structural facilities, costs of installation and management of the system components, and the impact that the collection process has on the characteristics of the waste (USDA 1999; see Chapter 9).

One major source water protection measure that can be done with some dairy and other operations is to eliminate the barnyard. If possible, open barnyards may be replaced with a concrete pad to better collect waste, and a roof to keep precipitation from contacting the manure. This will reduce not only runoff of manure constituents, but also soil erosion. Another important measure is the collection and isolation of calf wastes separate from that of other animals, since calves are a primary source of *Cryptosporidium* oocysts.

Waste Storage Facilities

Long-term storage of animal waste intended for land application is important for allowing it to be held until periods when climate, weather, and soil conditions, as well as plant requirements, are more optimal. Storage allows the farmer flexibility for performing land application at times when these conditions are best for coinciding with soil and crop needs and for protecting water sources (USEPA 2001a), and for when it may be more convenient for the task. The waste storage component of a farm's management plan should identify the required storage volume; storage period; the type, size, location, and installation cost of the storage facility; management costs for the process, and any impacts of the storage on the consistency or other characteristics of the waste (USDA 1999), as well as detailing the conditions when release from storage is allowed.

Different types of storage facilities of various sizes are used for different applications, including concrete flooring, shelters, holding tanks, pits, lagoons, and others. Some storage, such as liquid waste lagoons or solid waste compost piles, can also provide some degree of treatment. Storage alone is sometimes accomplished in pits located directly below the production facility (a practice common for hogs in cold climates and older egg-production facilities) or in either earthen storage pits or above-ground tanks, which are separated from the production facility.

As an example, poultry litter storage facilities are commonly used to keep rainwater and runoff away from poultry house waste. Various types of litter storage facilities are available, including (listed in order of most to least effective for pollution prevention) roofed structures, bunker-type storage, covered stockpiles, and uncovered stockpiles (USEPA 2001a). Clearly, covered facilities are the best for stopping precipitation from mixing with stored waste piles. Concrete flooring is often used to simplify waste collection procedures and restrict potential contamination into underlying groundwater. The LPES Curriculum provides a series of articles on manure storage at www.lpes.org/les_plans.html.

Waste Composting

Composting involves decomposition and stabilization of organic manure materials (USEPA 2001a), and is a common and relatively inexpensive method of handling livestock manure. This is a natural process that converts animal wastes, bedding, and other raw products into humus, and is enhanced and accelerated by the mixing of organic waste with other ingredients in a prescribed manner for optimum microbial growth (USDA 1999).

Composting manure reduces many of the typical drawbacks associated with raw manure use. Well-composted material is typically odorless, fine-textured, and low in moisture, and can be bagged and sold for use as organic matter with little odor or fly breeding potential. The availability of phosphorus, potassium, and micronutrients is greater in most compost than in raw manure, while nitrogen in compost is not as readily available (www.arb.ca.gov/ag/caf/dairypnl/compovrvw.pdf).

Aerobic composting reduces the mass and volume of the waste, reduces odors, can inactivate pathogens and kill weed seeds, stabilizes manure for easier handling, converts inorganic ammonia to a more stable organic form of nitrogen (stable humus has virtually no free ammonia or soluble nitrate), and can result in a valuable soil amendment (USDA 1999). Aerobic composting generates carbon dioxide, water vapor, and heat. Temperatures often reach 40 to 65°C during the active composting phase, and are capable of inactivating or killing assorted pathogens and parasites as well as weed seeds. As an example, in a study of composting beef cattle manure, Van Herk et al. (2004) showed that the viability of these pathogens decreases rapidly once temperature exceeds 55 °C, and they concluded that the exposure of *Cryptosporidium* oocysts and *Giardia* cysts to temperatures > 55 °C for a period of 15 days during composting appears to be an effective method for inactivation of these pathogens.

Anaerobic composting is used less frequently, as it is a less rapid process than aerobic composting, is not nearly as good at eliminating odors (odor production is a primary reason anaerobic composting is not frequently used), and does not result in high temperatures like aerobic composting. Anaerobic composting produces methane and carbon dioxide, as well as ammonia, organic acids, and hydrogen sulfide.

Compost sites should be covered if possible to prevent precipitation from soaking the piles, and if not, then rainfall runoff should be contained and controlled. Sites should not be located near drinking water wells and surface waters, and should be on relatively flat ground where water does not run off or collect (USEPA 2001a). Proper carbon-nitrogen ratios need to be maintained, so organic material (carbon) is often added to the waste piles. Appropriate moisture content is also important for optimal composting. Aerobic compost piles need to be turned periodically to provide oxygen (or provided some other method of aeration). Aerobic composting should take place at the correct temperature, and pathogens can be killed if an appropriate length of time is used.

Manure composting guidelines may be found in numerous sources, including the following available on the internet:

- USDA guidance (Chapter 10 of USDA 1999) www.wcc.nrcs.usda.gov/awm/awmfh.html
- Farm-Scale Composting Resource List - contains a large number of web links to available resources (National Sustainable Agriculture Information Service 2006) www.attra.org/attra-pub/farmcompost.html
- Manure Composting Manual (Government of Alberta, 2004) www1.agric.gov.ab.ca/$department/deptdocs.nsf/all/agdex8875?opendocument

- Guide to Compost and Organics (Klickitat County, 2005) – includes assorted fact sheets for system selection, design, construction, maintenance and troubleshooting www.klickitatcounty.org/SolidWaste/default.asp?fCategoryID Selected=965105457
- Field Guide to On-Farm Composting (NRAES 1999) www.nraes.org/publications/nraes114.html
- On-Farm Composting Handbook (NRAES 1992) www.nraes.org/publications/nraes54.html

Waste Storage and Treatment Lagoons

Lagoons and waste storage ponds are typically made by excavating earth and constructing berms to provide storage of animal waste that is primarily liquid in consistency. This storage capability is important for allowing the waste to be held until later periods when climate, weather, and soil conditions for land application, as well as plant requirements, are more optimal.

Lagoons have three distinct zones containing liquids, sludge, and solids. The liquid supernatant from the lagoons is often spread on cropland and pastures as fertilizer, and the sludge and solids are removed and disposed of less frequently. The potential water quality risks from lagoons include leakage into groundwater or surface water, overflow as a result of exceeding capacity or because of excessive precipitation, and, in the event of a major berm breakage, the release of a large volume of liquid waste to nearby streams.

Lagoons provide treatment in addition to storage, and can reduce the amount of organics, pathogens, nutrients and odors in the waste (USEPA 2001a). Nitrogen can be volatilized or denitrified and lost to the atmosphere, and some phosphorus will precipitate down into the sludge. As an example application, anaerobic lagoons are commonly used for liquid swine waste in warmer climates, with the supernatant being used as recycled water for flushing the waste collection systems, with ultimate land application of the supernatant as fertilizer. Plans need to be made to recycle or dispose of the sludge, and this may occur as often as every five years.

Design considerations for both anaerobic lagoons and aerobic lagoons (oxidation ponds) are discussed in Chapter 10 of the USDA Agricultural Waste Management Field Handbook (USDA 1999). Proper planning, design, and maintenance of waste lagoons are essential, as lagoons can contaminate groundwater if they are not constructed and maintained properly. Containment berm integrity and proper liquid freeboard levels must be maintained to prevent large-scale spills of liquid waste to nearby surface waters. Lagoons should be located in accordance with regulatory requirements for separation from water sources such as drinking water wells and surface waters. They should not be located in flood plains, and should be downgradient from drinking water wells. Lagoons and waste storage ponds should be constructed with a low-permeability liner; options include a synthetic material (e.g., geotextiles) or compacted clay or other soil material (typically an 18-inch layer of clay with a hydraulic conductivity of 10^{-7} cm/s or less) (USEPA 2001a).

Erosion of earthen containment systems can lead to seepage and subsequent contamination of groundwater, and to weakening of berm walls. Appropriate vegetative cover helps to maintain earthen berm integrity. Maintenance of liner and berm integrity is also important, especially during activities that could cause damage such as the solid waste removal process. Two practices to protect the liner are building a concrete access ramp for waste removal equipment, and operating equipment under dry conditions by first removing all the liquids and letting the solids dry (USEPA 2001a).

Maintaining an effective freeboard level in lagoons and storage ponds is important for ensuring they won't overflow during precipitation events. Lagoons should be built to include capacity for some long-term (i.e., permanent) storage of waste, plus a period (e.g., six months) of generated waste storage capacity, plus the volume of storage for a 25-year, 24-hour storm event (a 25-year storm is one that has a one-in-25 chance of occurrence in a given year), plus a freeboard safety factor (e.g., 20 inches). The LPES Curriculum includes a fact sheet on liquid level markers for uncovered manure storages and lagoons (www.lpes.org/cafo/15FS_Marker.pdf).

Breaching of liquid waste storage facility containment walls can lead to large volumes of liquid wastes being discharged directly into nearby surface waters. Earthen containment walls are most susceptible for breaching during periods of heavy rains, which also help to increase the liquid level in the lagoon. As an example, the retaining wall of an 8-acre lagoon at Oceanview Farms in Onslow County, North Carolina broke in June 1995 and spilled approximately 22-million gallons of liquid manure waste into the New River (Agricultural Animal Waste Task Force 1996). Lagoon containment walls built out of wood, metal, or concrete have also failed.

Land Application of Solid and Liquid Animal Wastes

Land application of livestock manures for plant uptake and crop production, is a traditional practice and proven technique for recycling the nutrients, and most animal waste produced in the US ultimately is applied to crop and pasture land as fertilizer. Proper land application of manures is an important component of managing animal wastes at many facilities. However, some operators have historically viewed the land application of manure as more of a waste disposal system than a nutrient management system. Application to fields has therefore often been dictated not by seasonal crop nutrient demands but by other factors such as manure storage capacity and the capacity and layout of the irrigation/disposal system (Harter, Meyer, and Mathews 2002). Of those farms that do have nutrient management plans, there is a wide range of levels of compliance with the plans.

Nutrient management plans should be prepared and implemented to specify the timing and location of land application of manure nutrients at agronomic rates that do not adversely affect source water quality, and require maintenance of appropriate records. In many cases the nutrient management plans and manure land application schedules are prepared by certified planners (consultants), and often are completed with assistance from the USDA-NRCS, Soil and Water Conservation Districts, and other similar organizations.

Nutrient analysis of soils is used to determine the nutrients already available for plant needs, and analysis of the manure waste provides information on its nutrient content. This information, combined with information on expected crop yields and nutrient requirements, is used to develop fertilizer recommendations for applying the needed amounts of nutrients without applying excess. Applying animal waste at the time of maximum nutrient uptake by the crops will help reduce the release of nutrients to surface runoff and percolation into groundwater, decrease the amount of manure needed to fertilize crops (thus saving fertilizer costs for the farmer), and also maximize the yield of the crops. A complete plan will include identification of the fields to be used, design specifications for the waste delivery system (including selection of appropriate necessary equipment), determination of application rates, volumes and locations, analysis of the value of the recycled waste, standard operating procedures, labor requirements, and equipment maintenance schedules (USDA 1999). Crop nutrient analyses should also be part of any monitoring program to determine the amount of nutrients removed from the cropping area.

Delivery of the manure to the root zone can enhance plant nutrient uptake and minimize residual losses to the environment. If possible animal wastes should not be applied to frozen or snow-covered ground, and especially not to saturated ground. The degree of flexibility for timing of land spreading waste in accordance with these criteria is largely dependent on available storage capacity as well as climate and weather conditions. When possible, it is good to till in manure, especially when land applied in the fall in colder climates. Periodic calibration of solid and liquid manure spreading equipment is also important for ensuring that the correct amount of manure is applied.

In some geographic areas, the amount of animal waste produced is greater than the amount that can be utilized by the crops in the area, at least without overdosing specific nutrients (such as phosphorus). In such cases, alternative uses for the manure may be explored, and programs may be developed to help move excess manure out of the source water protection area to farms that can use the manure. Such a program in the Eucha/Spavinaw watershed in Arkansas and Oklahoma (see case study description in Chapter 7) matches the needs of poultry growers with farms outside of the watershed that can use the chicken litter as crop fertilizer. This program has helped reduce the excessive amount of manure that was being spread within the watershed and causing water quality problems in the drinking water source reservoirs.

Numerous sources of useful information on nutrient management and manure land application are listed in Appendix A. A few of these references include the following:

- AFO Virtual Information Center for Nutrient Management (USEPA website) – includes links to numerous sources of information on nutrient management and manure land application (http://cfpub.epa.gov/npdes/afo/afoinfo.cfm?view =category&link_cat=25)
- Managing Manure Nutrients at Concentrated Animal Feeding Operations (USEPA 2004a) - technical guidance implementing CAFO Rule requirements, including land application methodology (http://www.epa.gov/npdes/pubs/ cafo_manure_guidance.pdf or http://cfpub.epa.gov/npdes/docs.cfm?program_id= 7&view=allprog&sort=name)
- USDA-NRCS Agricultural Waste Management Field Handbook (USDA 1999) includes guidance on land application of manure in Chapter 11 (http://www.info.usda.gov/CED/ftp/ CED/neh651-ch11.pdf)
- NRCS nutrient and pest management info - nutrient management tools and info http://www.nrcs.usda.gov/technical/nutrient.html
- CNMP Watch - is a good source for information designed to help AFOs and other agricultural producers with the process of developing and writing of CNMPs. Includes summaries of state CAFO programs. http://www.cnmpwatch.com/
- NRCS CNMP Technical Guidance - handbook with guidance for Comprehensive Nutrient Management Planning http://www.nrcs.usda.gov/programs/afo/cnmp_ guide_index.html
- The Livestock and Poultry Environmental Stewardship (LPES) Curriculum provides a series of articles on land application and nutrient management http://www.lpes.org/les_plans.html and http://www.lpes.org/CAFO_PDF_ Land. html
- Purdue Animal Manure Solutions - Lots of web links, tools and articles on manure management http://danpatch.ecn.purdue.edu/~epados/pams
- Ohio State University - lots of links to other related websites (http://www.ag. ohio-state.edu/~farmnet/links/nutmgt.html)

- Ohio Livestock Manure and Wastewater Management Guide – descriptions of a wide variety of measures http://ohioline.osu.edu/b604/
- Best Management Practices: Land Application of Animal Manure - 10-page report from Ohio State University Extension http://ohioline.osu.edu/agf-fact/0208.html

Alternative Uses for Manure

Manure is not simply an animal waste product, but a potentially valuable commodity. Animal manure has been land applied as a crop fertilizer throughout history. However, new technologies have also made it possible to convert manure into sellable products and as a useable energy source. This can be an especially attractive alternative in areas where land application of manure is limited for one reason or another.

Value-added products generated from animal manure can include composted and odor-free fertilizers, soil amendments or feed additives. Energy can be obtained from manure through either combustion, biological conversion (anaerobic digestion), or chemical conversion (to methanol). Descriptions of alternative uses and technologies for manure are available from the USEPA (USEPA 2002b) and USDA (Hannawald 1999). The USEPA report also includes a listing of organizations that can assist with alternative manure utilization technologies.

Vegetative Filter Strips

Vegetative filter strips (also called riparian buffers or vegetative buffer strips) are a type of BMP designed to trap and reduce sediment, nutrients (nitrogen and phosphorus), and microbial contaminants in stormwater runoff. This is one of the most common BMPs employed (along with fencing cattle out of streams), though some question their effectiveness, especially in saturated soils and in the winter (Wright and Gooch 2005). Vegetative filter strips consist of bands or areas of planted or indigenous vegetation located between a waste source and downgradient receiving water. They can remove pollutants via filtration, deposition, infiltration, absorption, volatilization, plant uptake, and decomposition (Harner and Kalita 1999). Near surface vegetation can increase soil organic content and local microbial populations, enhance porosity and soil permeability, slow down near-surface flow velocities, and provide more organic-rich surfaces for adsorption.

Field experiments show that vegetative buffer strips can significantly reduce nutrients and fecal coliforms from feedlot runoff (Keaton et al. 1998, Kalita and Harner 1998, Williamson et al. 1999). Research on the ability of vegetative buffer strips to remove *Cryptosporidium parvum* oocysts during surface and shallow subsurface flow demonstrated 1 to 3-log removal by filter strips under simulated rainfall conditions (Atwill et al. 2002). Vegetative filter strips can be effective at removing *Cryptosporidium* oocysts from surface runoff, however, viruses (PRD1 bacteriophage) and to a lesser extent bacteria (*E. coli*) are more easily transported (CRC 2004). Laboratory experiments have shown that a one-meter width (i.e., about three feet) of grass vegetation could reduce *Cryptosporidium* in surface water runoff by up to 99.99 percent compared to bare soil alone (CRC 2004). The *Journal of the American Water Resources Association* devoted a whole issue (Vol. 42, No. 1, February 2006) to scientific papers presented at a 2004 conference related to the role of riparian zones and other buffers in the maintenance and improvement of water quality in agricultural areas.

Factors to consider during design of vegetative filter strips include the amount and characteristics of the potential runoff waste, slope of the land, lateral dimensions of the vegetated area, soil type, selection of vegetation type and density, and other factors. Further information about vegetative buffers is available from the following sources:

- NRCS Buffer Strips: Common Sense Conservation – good website with information about buffer strips and links for more detailed reports and guidance (www.nrcs.usda.gov/feature/buffers/)
- NRCS Economics of Conservation Buffers - web page with links to information sources on buffer economics (http://www.nrcs.usda.gov/technical/RESS/issues/bufferecon.html)
- Evaluation of Riparian Buffers and Controlled Drainage – summary of a North Carolina State University research project (www.epa.gov/region04/water/nps/projects/nc97-1.htm)

Constructed Wetlands

Constructed wetlands can be used for the management of diluted liquid animal waste, including dairy, swine and poultry, by providing a cost-effective waste treatment system. Constructed wetlands effectively remove pathogenic microorganisms through various processes. Viruses can be adsorbed by soil and organic particles while bacteria are removed or inactivated by sedimentation, ultraviolet radiation, chemical reactions, natural die-off, and predation by zooplankton. One study of wetlands used for swine wastewater treatment showed they were effective in removing over 75 percent of total nitrogen and ammonia nitrogen, but were not as effective at removing phosphorus (Stone et al. 2000). Further information about constructed wetlands for animal waste treatment is available from the following sources:

- Fact sheets on constructed wetlands for animal waste treatment:
  - Alabama Cooperative Extension System - includes summaries of research on using constructed wetlands for treating dairy and swine waste (Tyson 1996) (www.aces.edu/pubs/docs/A/ANR-0965/)
  - North Carolina Cooperative Extension Service (Humenik et al. 1996) www.bae.ncsu.edu/programs/extension/publicat/wqwm/ag473_13.html
  - Ohio State University Extension (Simeral 1998) http://ohioline.osu.edu/a-fact/ 0005.html
  - Purdue University Cooperative Extension Service (Miller, MacGowan, and Reaves, 2003) www.ces.purdue.edu/extmedia/FNR/FNR-202.pdf
- USEPA guide for 'Constructed Wetlands for Animal Waste Treatment: A Manual on Performance Design and Operation with Case Histories' (USEPA 1997a) (http://yosemite.epa.gov/water/owrccatalog.nsf/0/caf5289c155133ef85256d83004fd7e0?OpenDocument)
- USEPA guide for 'Constructed Wetlands And Wastewater Management For Confined Animal Feeding Operations' (USEPA 1997b) (http://yosemite.epa.gov/water/owrccatalog.nsf/0/676abe7d693e905585256d83004fd9c1?OpenDocument)

Irrigation Practices

Large amounts of water are used in many irrigation systems, and, similar to rainfall events, these systems provide substantial potential for transport to receiving waters of contaminants that have been land applied. Irrigation water should be managed and used to optimize the amount of water applied and maximize the efficiency of where and when it is applied in relation to crop needs (and thus saving costs on water). This will help to minimize runoff and reduce the amount of excess water leaching contaminants into the soil. Watering systems such as sprinklers, low-energy precision applications, surges, and drips help deliver water uniformly and efficiently. Instead of natural earthen water transmission systems, efficiency can be improved by using lined ditches and gated pipe. In some circumstances, field drainage recovery ponds can be used to capture irrigation water for reuse as well as capturing sediments and nutrients (USEPA 2001b).

Other Treatment Systems for Animal Waste

Treatment may include physical, biological, and chemical processes, and is designed to reduce the pollution potential of the waste (e.g., aerobic and anaerobic digestion, rotating biological contactors) or change the solids content for handling purposes (e.g., mechanical separation, settling basins, and drying). Treatment systems can be designed for removal of the contaminants of concern, just as with conventional municipal and industrial water and wastewater treatment, and can be used to stabilize organic matter and nutrients (N and P), control pathogens, reduce odor and gas, or improve handling characteristics during storage and spreading. As an example, anaerobic digesters are used at some facilities for manure treatment with good success (e.g., Moser 2006; www.epa.gov/agstar/resources.html). Some farms use treatment systems to help deal with milk house waste or egg wash water. Proper operation and maintenance of these facilities is essential for achieving the desired treatment results.

Stabilization of nutrients by alum (aluminum sulfate) is a relatively new area of research (USEPA 2004b), and an area where a water utility can use their own treatment plant's residuals to help in source water protection from agricultural operations. Application of alum and ferric hydroxide water treatment residuals to agricultural land can provide an inexpensive disposal option for utilities and help in source water protection by binding to phosphorus and reducing phosphorus solubilization after land application of manure and litter (DeWolfe 2003, DeWolfe 2006, Walker 2000). Alum applied to dry litter in poultry houses also reduces ammonia gas emissions from the litter, which in turn improves bird health, increases weight gains and reduces mortality, thus saving the producer money. In fact, savings from these benefits may more than offset the cost of the alum (Walker 2000).

Cost and maintenance considerations are usually prime factors when considering more advanced treatment technologies versus conventional land spreading of manure. Often advanced treatment is seriously considered only when substantial water pollution risks or odor problems are identified in nutrient management plans. Summaries and relative evaluations of potential treatment methods for AFO wastes are provided by the following references:

- USEPA – Chapter 8 on 'Treatment Technologies and Best Management Practices' in the Development Document for the Proposed Revisions to the NPDES Regulation and the Effluent Guidelines for CAFOs (USEPA 2002c)
- USDA Agricultural Waste Management Field Handbook – see Chapter 10 for descriptions of design criteria for numerous animal waste treatment and handling methodologies (USDA 1999)
- The LPES Curriculum provides articles on treatment of animal wastes at www.lpes.org/Lessons/Lesson25/25_Manure_Treatment.html and www.lpes.org/Lessons/Lesson43/43_4_Biological_Processes.pdf

Dead Animal Disposal

All animal farming facilities produce animal mortality, and animal carcasses need to be properly disposed. As an example, large poultry operations experience mortality almost daily, with highest mortality occurring when the chicks are young and during their last week or two where heart attacks are spurred by the relatively high growth rate of the chickens. Dead animals must be removed from the housing facilities as soon as they are found to prevent spread of disease. AFOs should consult with the appropriate authorities to help comply with applicable state and local laws about animal mortality disposal. Farms should have contingency plans prepared for disposal of large quantities of animals in the event of a major incident causing large-scale mortality.

Common techniques for disposal of dead animals include composting, rendering, incineration, burial, and landfilling. Rendering is where the dead animals are taken to a facility where they are ground up for use in animal feed, as a base material for cosmetic products, and other uses. Rendering is a very good option where possible, as the farmer simply needs to collect and store the dead animals for pick up by a third party, and the carcasses are transformed to a useful saleable product. Unfortunately, rendering services are not available to many producers. Pick up service for rendering should be within a day or two of death. On-site storage facilities such as a dumpster (ADU container, or animal disposal unit container) should be used to hold the dead animals waiting to be picked up for disposal. Some facilities use freezers for temporary storage of animal corpses. The second preferred option is composting, which also results in a usable product. Burial, landfilling, or incineration can also be used, though they may not be legal in some localities. Where allowed, common burial guidelines include locating grave sites at least 200 feet away from any streams or other waterbodies, and to bury under a minimum of three feet of well-packed soil (which can be challenging when ground is frozen. For more information on dead animal disposal practices, see USDA (1999) and LPES Lesson #51 on Mortality Management (www.lpes.org/Lessons/Lesson51/51_Mortality_Management.html).

Fuel Spill Prevention

Farm operations usually include the storing of fuel oil, diesel, gasoline, and other fuels and chemicals. Spill prevention and control measures should be included as part of the farm management plan, including procedures for the use and filling of the storage containers. Fuel and chemical storage should be located to minimize the potential for pollution of nearby wellheads or surface waters in the event of a leak or spill. Storage structures should be constructed of materials that are chemically compatible with the substances stored. Secondary containment structures should be designed to hold at least 110 percent of the volume of the largest container. Spill Prevention, Control, and Countermeasures (SPCC) plans may be required in some cases (see www.epa.gov/oilspill/spcc.htm).

CHAPTER 6
STRATEGIES AVAILABLE TO WATER UTILITIES

In addition to working with AFO operators and owners and the appropriate regulators to encourage employment of BMPs to control contaminants at the source, utilities should understand and examine other alternatives available for them to deal with the contaminant releases that occur. This includes considerations for chronic contaminant releases as well as contingencies associated with acute accidental releases. Strategies that utilities can themselves employ include the following:

- Source water assessments
- Identification of the type and location of CAFOs in a source water area
- Source water protection activities
- Cooperation with stakeholders
- Education for stakeholders
- Funding alternatives
- Source water monitoring
- Contaminant source tracking
- Early warning monitoring
- Management and timing of alternate intake locations
- Pre-sedimentation
- Off-stream storage
- Riverbank filtration
- Water treatment processes

Several case studies that have successfully used these approaches are presented in Chapter 7.

SOURCE WATER ASSESSMENT

As a first step, water utilities need to assess which potential waste sources may have the greatest potential adverse impact on source water quality, and prioritize protection activities and measures for action in their source water area. This analysis should include all potential waste sources, not just those from the agricultural sector or CAFO facilities alone. For the agricultural sector, water utilities should be concerned not just about AFOs and CAFOs, but about poorly managed farms of any type. Periodic reassessment of the action priorities and the success of approaches taken will help ensure the program is achieving the optimal benefits.

All source waters should have had a Source Water Assessment Program (SWAP) report prepared under requirements of the 1996 SDWA. If a water utility is concerned that the SWAP for their source water was not thorough enough in identifying potential contaminant sources and prioritizing their potential impact, then they are encouraged to conduct additional study and monitoring. Advanced studies use hydraulic and contaminant transport modeling of watersheds or groundwater to more fully estimate potential impacts from priority contaminant sources. Modeling can also be used to assess the anticipated general effect of possible BMPs and to prioritize actions. Many utilities have conducted this type of thorough analyses of the potential impact of the actual and potential priority pollution sources in their source water area (e.g.,

Tulsa, Oklahoma and Philadelphia, Pennsylvania, as discussed in the case studies presented in Chapter 7).

A sound source water assessment that analyzes the relative contributions of potential pollutant sources in a source water area can go a long way toward providing the scientific evidence necessary to effectuate change in land use practices. Utilities can help themselves by bringing in the right experts and expertise to perform sound scientific assessments. Data and sound science showing agricultural impacts on source water quality should be locally and regionally based to be more convincing to local and regional agricultural interests.

IDENTIFICATION OF CAFO TYPES AND LOCATIONS IN A SOURCE WATER AREA

Some water utilities already know about the type, size, and location of CAFO, AFO, and other agricultural operations in their source water area through the SWAP report or other studies. However, many water utilities do not know if there are AFOs or CAFOs in their area, or how many there are and of what type. No single database was found during this research that could be used by a utility to identify these sites. Instead, a utility can contact the appropriate federal, state or local agencies (such as the state primacy agency or regional USEPA office), conservation districts, cooperative extension offices, and industry associations to gather this information for a specific area. A good place to start would be with the CAFO Rule primacy agency in your area. General data on animal production on a county basis are also available from the USDA, which can give an initial indication of the number of AFOs in an area. Working with these agencies will help to define the agricultural makeup of a specific watershed. Some suggestions of contacts are provided below:

- Links to state environmental agencies responsible for administering AFO and CAFO regulatory programs http://cfpub.epa.gov/npdes/afo/afoinfo.cfm?view=category&link_cat=24
- List of state environmental agency primary personnel contacts for AFO/NPDES issues http://cfpub.epa.gov/npdes/contacts.cfm?program_id=7&type=STATE
- List of conservation district offices in each state (National Association of Conservation Districts) www.nacdnet.org/resources/cdsonweb.html
- List of state conservation agencies (members of National Organization of State Conservation Agencies) http://nascanet.org/agencies.htm
- Industry associations such as for poultry (list available at www.wvu.edu/~agexten/poultry/assoc.htm) or cattle (e.g., Texas Cattle Feeders Association; www.tcfa.org)
- USDA 2002 Census of Agriculture (USDA 2004) – Provides data on the number of farms for different animal types, as well as estimated production and sales, for each county and state in the US
  - written report with state-level summary data available at www.nass.usda.gov/census/census02/volume1/us/USVolume104.pdf
  - searchable database for county- or state-level data at www.nass.usda.gov/Census_of_Agriculture
  - searchable database also at http://agcensus.mannlib.cornell.edu/
- USDA mapping - An on-line interactive map is available that draws from USDA Census of Agriculture database and can be used to identify geographic areas with

various amounts of animals or animal farms is available at www.nass.usda.gov/Census_ of_Agriculture/SVG/index.asp

SOURCE WATER PROTECTION

Reduction of contaminant emissions and use of BMPs at the AFO source (as described in Chapter 5) are proactive and desirable solutions from the perspective of the water utility because it is always beneficial to deal with contaminants at the source rather than trying to deal with them in the environment or after the drinking water intake. Water utilities can use Clean Water Act programs to help with source water protection to meet their Safe Drinking Water Act responsibilities. The Total Maximum Daily Load (TMDL) program can be an effective avenue for watershed pollution control, and water utilities should be proactive in development of TMDLs in their watershed. Using TMDLs helps to provide a good scientific basis for recommending pollution reduction BMPs, and also can help to increase stakeholder involvement. Clean Water Act Section 319 grant funds are often used for environmental assessment, protection, restoration and education projects.

Guidance for planning and implementing source water protection programs is available from a wide variety of sources, including the following (see Table 5.3 for references specifically for controlling animal and agricultural wastes):

- USEPA Source Water Protection webpage (www.epa.gov/safewater/protect.html) contains links to a variety of guidance materials
- USEPA Annotated Bibliography of Source Water Materials (USEPA 2003f).
- USEPA draft Handbook for Developing Watershed Plans to Restore and Protect our Waters (USEPA 2005d)
- Introduction to EPA's Drinking Water Source Protection Programs (USEPA 2003e)
- Source Water Protection: Best Management Practices and Other Measures for Protecting Drinking Water Supplies (USEPA 2002d)
- State Source Water Assessment and Protection Programs Guidance: Final Guidance (USEPA 1997c)
- Protecting and Restoring America's Watersheds: Status, Trends, and Initiatives in Watershed Management (USEPA 2001c)
- Getting in Step: A Guide for Conducting Watershed Outreach Campaigns (Tetra Tech, Inc. 2003)
- USEPA Watershed Academy On-Line Training Modules (available at www.epa.gov/watertrain/, e.g., www.epa.gov/watertrain/pdf/swp.pdf (January 2003) and www.epa.gov/watertrain/pdf/swpbmp.pdf (August 2002)).
- USEPA National Agriculture Compliance Assistance Center website - information on animal production practices and BMPs (www.epa.gov/agriculture)
- USEPA Animal Feeding Operations (AFO) Virtual Information Center website (http://cfpub.epa.gov/npdes/afo/virtualcenter.cfm)
- AwwaRF Source Water Protection Reference Manual (CDM 2002)
- Source Water Protection: Effective Tools and Techniques You Can Use (1999 Participant Manual) (AWWA 1999)
- Effective Watershed Management for Surface Water Supplies (Robbins et al. 1991)

- Guidance to Utilities on Building Alliances with Watershed Stakeholders (Raucher and Goldstein 2001)
- Protecting the Source: Land Conservation and the Future of America's Drinking Water (Ernst 2004)
- Path to Protection: Ten Strategies for Successful Source Water Protection (Ernst and Hart 2005)
- Source Protection Handbook: Using Conservation to Protect Drinking Water Supplies (Hopper and Ernst 2005)
- Source Protection: A National Guidance Manual for Surface Water Supplies (NEIWPCC 2000)
- The Center for Watershed Protection provides basic templates to help with design of watershed protection programs; see www.stormwatercenter.net.
- From Source to Tap: Guidance on the Multi-Barrier Approach to Safe Drinking Water (CCME 2002)
- Source Water Assessment & Protection Workshop Guide, Second Edition. (Herpel 2004; Groundwater Foundation, www.groundwater.org)

COOPERATION WITH STAKEHOLDERS

Perhaps the most important step for source water protection in this case is to develop a dialogue with the AFOs and CAFOs of concern. Many CAFO operators want to be "good neighbors" and have no desire to pollute the watershed. Such proactive CAFOs may be willing to work with regulators and drinking water utilities to develop mutually beneficial partnerships. Working in harmony, reasonable communication trees can be established and sound, cost-effective control measures put in place.

Establishing good communication and collaboration between the water utility, farmers, regulators, and other key stakeholders is essential for effective source water protection, and can not be overemphasized. One of the most effective means of approaching agricultural interests is through cooperation with entities that already have relationships established with the local farmers. These can include county soil and water conservation districts, USDA NRCS offices, USDA Cooperative State Research, Education, and Extension Service (CSREES), land grant universities and their associated Cooperative Extension Services, Farm Bureau, or agricultural industry organizations. Other state and federal agencies may also be willing to provide financial and technical assistance to help improve water quality. Water utilities should take advantage of these resources, rather than try to do all the work on their own. This is both a more effective and more economical approach.

Working through local agricultural and watershed stakeholder organizations and coalitions generally builds trust better than when utilities initially approach farms and the agricultural community on their own. Most successful cooperative efforts involve the utility first contacting an agricultural outreach organization (e.g., Soil and Water Conservation Districts, Cooperative Extension offices, and the NRCS) that the farmers already know and trust. Having these agricultural partners approach the landowners and agricultural interests is usually much more fruitful than when done directly by a 'big city' government located many miles downstream. The agricultural outreach professionals can establish the initial contact with the farmer, help bring the parties together, and serve as a mediating force.

Water utilities should start their approach to the local agricultural community very non-aggressively. Trust must be established before changes in farm operational practices are likely to be made. When utilities push too hard and too fast, they sometimes accelerate and intensify resistance. Many of the potential BMPs are actually management practices that farms need to adapt into their regular operations. BMPs, if imposed on farms, are often not likely to be operated and maintained to fully achieve the desired pollutant reductions. Getting the buy-in of the farmers regarding the benefits of the BMPs, both to the environment and to their operations, is important for them to feel a need to continue proper BMP maintenance. Monitoring every farm and farm operation is usually not a realistic method of control.

Communication with local interests is key. Every watershed is different, and utilities need to learn who the effective players are in their area. Utilities need to identify what lines of communications are most beneficial. Foremost, farmers listen to other farmers, and then to the cooperative extension and soil conservation groups. Producers in integrated animal industries must also follow the directions provided by their corporate integrator. In many cases animal farmers also get a lot of their information from professionals who visit their operations such as veterinarians and nutritionists.

Always try to find win-win scenarios with the local farmers. Approach the farms asking what you can do to help them, as opposed to telling the farmers what to do. Emphasize the benefits that both their farms and the environment will realize from the suggested improvements in farm operations. Also realize that the first question will usually be who will fund the suggested BMP improvements. Providing incentives such as funding for BMP implementation or identification of measures that can increase farm productivity and profitability can be very useful. Providing means for recognition of the farmers' efforts (e.g., environmental stewardship award programs, public recognition in the water utilities' consumer confidence report, etc.) can also be a good motivator.

The dairy and beef industries are decentralized, dairy in particular, while the poultry and swine industries are very centralized (integrated). As such, different approaches are needed for poultry and swine versus dairy and beef. Because of this difference, dairy and beef farmers may need to be approached more on an individual basis, though it is better to use entities that already have relationships established with the farmers (as previously discussed). Successful approaches for the centralized (integrated) swine and poultry farms should also be toward the local level, but will definitely need corporate-level buy-in. Animal industry associations can also be a good contact. For example, Texas Cattle Feeders Association, the largest cattle industry association in the US, covering Texas, Oklahoma, and New Mexico, is well familiar with beef CAFO operations in their area. Working with these types of associations can potentially be more effective than water utilities trying to directly approach individual producers.

A recent AwwaRF report on "Water Utility/Agricultural Alliances – Working together for Cleaner Water" addressed the benefits and means of establishing cooperative alliances between water utilities and agricultural interests (Fletcher, Davis, and Pyke 2004). The report identifies strategies for drinking water utilities to build successful source water protection alliances with agricultural producers at the local, state, and national level. It notes that to be successful, "alliances must identify the problems in a watershed and then pool resources from diverse partners to design solutions, seek appropriate funding and follow through with the solutions that ultimately improve water quality" (Fletcher, Davis, and Pyke 2004). Using sound science to identify the problems in the source water area and to prioritize all potential contaminant sources and solutions, as well as educational outreach of the methodology and results of the prioritization analysis, are crucial for gaining support of the agricultural community.

Fletcher, Davis, and Pyke (2004) cited five common factors that characterize successful source water protection alliances:

- Each successful alliance builds relationships with community partners and utilizes them to create an alliance structure based on the strengths, skills, goals and needs of alliance members
- These alliances prioritize goals by assessing the needs in their watersheds
- They execute their objectives by determining the strategies that work best in their community
- Each alliance takes action to improve water quality
- The follow-through and long-term commitment of alliance members is invaluable

Fletcher, Davis, and Pyke (2004) also outline the motivations that agricultural interests may have in joining and participating in such an alliance. In addition to many animal producers being conservationists and having a vested interest in protecting natural resources, by joining in an alliance with water utilities the producers may gain substantial financial, educational, and networking benefits. Examples of benefits that may be experienced by producers are cost-sharing for implementation of agricultural BMPs on their farms, opportunities to view BMP demonstration projects, technical assistance from alliance partners, increased profits (e.g., from saving commercial fertilizer costs by using a manure nutrient management plan that optimizes the use of manure fertilizer on forage crops), and networking opportunities with agricultural educators and other producers (Fletcher, Davis, and Pyke 2004).

The case studies provided in Chapter 7, as well as those described by Fletcher, Davis, and Pyke (2004), illustrate a variety of means used by water utilities to form cooperative relationships with local farmers and other agricultural interests. In some cases, it may be most effective for a water utility to work through an umbrella watershed organization (such as a Watershed Agricultural Council) with a large variety of stakeholder participants that addresses a variety of prioritized issues, including agricultural ones. The Schuylkill Action Network in the Philadelphia area presents one excellent example of this approach (see case study in Chapter 7) for a water utility that has AFOs and CAFOs located in their source water area but outside of their political jurisdiction. For integrated industries such as swine and poultry, it can sometimes be more effective to work through the integrators instead of the individual farms. The integrators often control the methods used at these farms, and can thus exert pressure for changes in operating procedures. In addition, integrators provide technical assistance on their contract farms and are in a good position to disseminate information.

STAKEHOLDER EDUCATION

Education of AFO and CAFO operators is often essential for good cooperation and progress toward protecting source water quality. Many people have a misperception that water treatment plants can remove all of the pollutants from water, and also do not understand the impacts their activities can have on surface water and ground water quality. Educational outreach to the local agricultural community can help overcome these factors, and lead to improved cooperation among the various parties. A lot of these science and environmental issues may be new to some farmers, and accepting there is a need for change in their practices can take time. Farm operators will need to be assured and educated, usually through sound science, that their practices contribute substantially to the source water quality problems experienced by the utility, and that changes made in their practices will result in protection or improvement of source water quality. Once a producer understands the impact of their activities, and affordable and effective solutions are identified, they are more likely to make needed changes to their operations.

Examples of some existing outreach and educational programs used by water utilities are provided in the case studies in Chapter 7, and include public education campaign methods such as workshops, publications, informational web sites, posted signs, print and broadcast media, environmental conservation awards, tours, and demonstration projects. Some of these activities are conducted by (and perhaps contracted to) organizations that are already involved in educating the agricultural sector at the local level, including Soil and Water Conservation Districts, Cooperative Extension Offices, and agricultural professional organizations. Since these parties are often already known by the producers, the level of trust already established can improve the results of an educational program. Farmer-to-farmer education can also be highly effective.

In areas with water utilities proactive in source water protection, individual education is sometimes conducted at the farmer's facilities, which allows for direct assessment and discussion about the practices conducted at that farm. Farmer involvement in development of their Whole Farm Plans and Nutrient Management Plans leads to a better understanding and acceptance of what they can and should do. Some water utilities, such as Tulsa (Oklahoma) and Springfield (Missouri), support demonstration farms that are used to showcase various agricultural BMPs, and educate farmers as to more beneficial approaches.

Education of regulatory agency personnel, and of agency or utility personnel who perform inspections of agricultural activities in the source water area, is also beneficial. Information about farming practices, the practical nature and limitations of those practices, and the economics of the business, can help them better understand the farmer's situation and position, and lead to improved communication between the parties. On-site inspections conducted by agency or utility personnel can also be good avenues for mutual education, and education of both the farmers and the inspectors is important for maximizing cooperation between the two parties.

Training modules and programs, as well as guidance for development and implementation of environmental education programs, are available on the USEPA Source Water Protection website (www.epa.gov/ogwdw/protect.html) and other sources such as www.epa.gov/water/education.html and many of the on-line references listed in Appendix A.

FUNDING ALTERNATIVES

Funding is a common obstacle to improvement of agricultural operations for protecting drinking water source quality. Farmers often are not willing to implement non-required BMPs that adversely affect their ability to cover expenses and make a profit. Many successful case studies have shown some type of cost-sharing to be effective. Of course, farmer cooperation is best when funding sources are available to pay for 100 percent of the costs of BMP construction and implementation. Proactive water utilities can work with farmers and other stakeholders to find funding for a variety of agricultural source water protection projects. Knowledge of these potential funding sources, as well as good relationships with stakeholder organizations that are experienced at gaining related funding, can greatly help a water utility to gain cooperation of the farmers in their source water area.

The USDA Natural Resources Conservation Service (NRCS) offers many cost-share programs to farmers to implement environmentally sound farming practices, such as the Environmental Quality Incentives Program (EQIP), the Agricultural Management Assistance (AMA) Program, and others. All NRCS cost-share grants require farms to have a CNMP, regardless of size. EQIP is the primary source of USDA financial assistance to AFO owners and operators, and was initiated in the 1996 Farm Bill. Information sheets describing these NRCS programs and funding available to CAFO owners and operators are available at www.nrcs.usda.gov/programs/afo/CAFO_AsstAvail.html and http://www.lpes.org/CAFO_PDF_Farm.html. These funds can be used for development of Comprehensive Nutrient Management Plans and implementation of BMPs. USDA also has webpages with links for federal funding opportunities for rural areas (www.nal.usda.gov/ric/ricpubs/funding/federalfund/ff.html) and for small farms (www.nal.usda.gov/ric/ricpubs/ small_farm_funding.htm). In addition, funds for pollution prevention measures may be available from the USDA-NRCS Conservation Reserve Program (CRP, www.nrcs.usda.gov/ programs/crp/), USDA-NRCS Cooperative Conservation Partnership Initiative (www.nrcs.usda.gov/programs/ccpi/index.html), and other related programs (see www.nrcs.usda.gov/programs).

USEPA is also a source of funding for projects that assess a variety of issues, including Federal Clean Water Act Section 319 grant funds for environmental assessment, protection, restoration and education projects. USEPA has summarized the federal financial (grants and loans) and technical assistance programs that are available for AFOs and CAFOs to address environmental concerns (USEPA 2005c). Descriptions of federal and state watershed, groundwater and source water protection funding programs are available at www.epa.gov/water/funding.html and www.epa.gov/safewater/protect/financial.html. An informative May 2003 USEPA report on federal funding sources for source water protection activities is available at www.epa.gov/safewater/protect/pdfs/guide_swp_swp_funding_matrix.pdf. USEPA also maintains an on-line searchable database of financial assistance sources (grants, loans, cost-sharing) available to fund a variety of watershed protection projects (http://cfpub.epa.gov/fedfund/). This Catalog of Federal Funding Sources for Watershed Protection allows searches by the type of organization seeking funding (e.g., water utilities, business, private landowner, state or federal agency, community/watershed organizations, non-profit groups, etc.), the type of assistance sought (grants or loans), and select key words (e.g., agriculture, best management practices, drinking water).

Individual states may also have similar programs, and state environmental agencies are also a good potential source of funding for watershed and agricultural projects. For example, Virginia has state tax credits that allow farmers to deduct a percentage (typically 30 percent) of the cost of selected BMPs from their state tax obligation. The credit can be taken over several years if it exceeds what is owed in a single tax year.

In some cases, a water utility may wish to purchase sensitive lands or acquire easements to take them out of agricultural production (and protect them from other development). There are several non-profit organizations that can assist in such efforts, including providing and acquiring funding (e.g., The Trust for Public Land, http://www.tpl.org). Education of the property owners can be conducted in conjunction with negotiations for easements.

Some water utilities will pay for a portion or all of the costs of agricultural BMPs implemented within their source water area. Some of those who pay 100 percent of the cost of implemented BMPs are utilities with very proactive source water protection programs developed as part of a Filtration Avoidance waiver (e.g., New York City and Syracuse, New York). These utilities may also seek outside funding sources to lessen their financial commitment. However, despite the attractiveness of having BMPs totally paid for by outside parties, sometimes in these cases the lack of a financial investment does not provide a sense of ownership of the BMP to the farmer, and thus these farms may at times be less motivated to implement the appropriate maintenance.

SOURCE WATER MONITORING

Monitoring of source water quality is important for identifying pollutants and their sources (to target protection efforts), and for verifying the effectiveness of source water protection programs (including monitoring the effectiveness of BMPs at AFOs). Many monitoring programs target both dry and wet (storm event) sampling of their source water and upstream locations, since the primary pollutant source may vary depending on the amount and recent frequency of rainfall and runoff, the amount of fixed discharges (e.g., public sewage treatment plants), and other factors. Monitoring programs are used as a primary intake water protection measure by many utilities that have no other means of controlling outside activities in their source water area, including utilities drawing source water from major rivers with extensive watersheds. Monitoring techniques for select parameters of interest are discussed below.

Microbiological Analyses

Microbiological analyses are used to identify and quantify bacteria, viruses, protozoa and other potentially harmful pathogenic microorganisms. Conventional methods involve three steps for detecting and identifying microorganisms: (1) recovery and concentration of the target species, (2) purification and separation of the organisms, and (3) detection and identification.

Even though relatively easy-to-identify indicator organisms are often used for indicating the potential presence of more harmful microbes, significant drawbacks to analytical methods for microbes still exist in terms of their applicability for early warning monitoring applications. The primary weakness of conventional methods of microbial analysis is the relatively long time period required for isolation and reproduction (amplification) of the microbiological species, often a day or more. In addition, many tests are specific only to a single species or class of organism, while raw waters can contain a wide variety of pathogens.

General sources of information on conventional methods for the detection and measurement of microbial species are found in USEPA methods (http://www.epa.gov/ogwdw000/methods/indmicro.html), Part 9000 of Standard Methods (APHA et al. 2005), Baker and Bovard (1996), and many other sources. Rose and Grimes (2001) reviewed recent developments in microbial analysis. A recent review of emerging pathogens also included discussion of analytical methods for their detection in water (AWWA Microbiological Contaminants Research Committee 1999a, 1999b). Rapid detection methods for microbes are summarized in Venter (2000) and in Section 9211 of Standard Methods (APHA, AWWA, and WEF 2005).

The microbiological quality of drinking water has for years been assessed through detection of indicator bacteria such as total coliforms, *E. coli*, or heterotrophic plate counts (HPC). Conventional methods involve cultural techniques that take 18 hours or more to perform. Recent advances in testing procedures have reduced the analysis time to the range of 4 to 8 hours or less (e.g., Prescott and Fricker 2002; Eccles, Fricker, and Fricker 1998; Lee and Deininger 1999; Deininger and Lee 1998). Commercial methods are often used that involve specific enzyme defined substrate technology for monitoring for total coliform and *E. coli* within 18 to 24 hours (e.g., Edberg et al. 1991, Edberg and Smith 1994).

Among the many protozoan parasites of concern to the drinking water industry are *Giardia* and *Cryptosporidium*. *Giardia* has been known for years to cause severe diarrhea and related problems in humans. *Cryptosporidium* is comparatively new to the water industry, and has caused several major outbreaks of Cryptosporidiosis over the last decade or so. Even though commercial kits are available that simplify the concentration step using immunomagnetic separation systems, the USEPA-approved methods (1622 and 1623) require extensive training for their immunofluorescence microscopy technique.

Turbidity

Turbidity is an indicator of water clarity as measured by the amount of light scattering caused by particles in water, and is an indication of the amount of suspended particles present in a sample. This parameter is regularly used in drinking water treatment process control applications, and testing of source waters is often performed for this purpose. Of substantial importance for surface water early warning applications is the frequent correlation between large increases in turbidity measurements and the potential for adverse changes in microbiological water quality. This relationship results from the fact that both turbidity and microbial concentrations often increase substantially during and subsequent to storm events. As rainwater flows overland into receiving streams and other waterbodies, it picks up both soil particles and microorganisms (microorganisms also frequently attach to soil particles). This can be especially problematic in agricultural areas where vegetative groundcover is low and significant amounts of microorganisms exist from animal husbandry operations.

As an example of the application of turbidity for an early warning indicator of increased pathogen concentrations, a New Jersey water utility investigated the correlation of storm events and increased flow, turbidity, and the presence and concentration of the protozoan parasite *Cryptosporidium* (LeChevallier et al. 1998). The increased flow rate during the spring rains largely comes from overland runoff, which carries with it soil particles and pathogens. The resulting increases in turbidity were accompanied by increases in *Cryptosporidium*. Analysis of the data showed that high *Cryptosporidium* loadings could often be avoided by shutting the intake when turbidity rises above a certain level (e.g., > ~15 ntu). The treatment process at the particular utility where this analysis was performed incorporates on-site raw water storage that allows the intake to be shut down for about 24 hours without interruption in service. As was done for the above example, use of a surrogate such as turbidity should be supported by evidence of a correlation with the parameter(s) of concern in the specific source water area.

On-Line Analytical Probes

On-line analytical probes, including ion-selective electrodes, are extremely convenient for continuous or nearly-continuous early warning monitoring applications, and are available from a variety of manufacturers as single probes or as sets that monitor multiple parameters. The latter devices typically use different ion-selective and other electrodes to measure levels of several parameters at once, including temperature, pH, turbidity, conductivity, dissolved oxygen, oxidation-reduction potential (ORP), nitrate, NH_3/NH_4, chlorophyll, select metals, and physical parameters such as water level.

Dissolved Oxygen

Dissolved oxygen is essential for the survival of aquatic life, and thus is almost always used in ecological monitoring programs. Oxygen probes can provide continuous and on-line measurements, and are relatively inexpensive. Sudden reductions in oxygen may be indicative of releases of organic constituents by wastewater collection and treatment systems, combined sewer overflows, and surface water runoff. Given that these releases also often include microbial pathogens and other important pollutants, measurement of source water dissolved oxygen can be a relatively easy, though not very specific or sensitive, means of monitoring gross changes in water quality from these sources. In addition, diurnal fluctuations in dissolved oxygen can be indicative of the presence of algae; thus oxygen levels are sometimes used in conjunction with chlorophyll and turbidity measurements to monitor for algal blooms. Most multi-parameter probes include units for measuring dissolved oxygen.

Nitrate and Ammonia

Nitrate and NH_3/NH_4 may be measured with a specific ion electrode, and instruments for on-line colorimetric and UV analyses are also available. However, the lesser expensive specific ion electrodes have not yet proven very effective for detecting changes at low concentrations typically found in surface water supplies.

General Organic Chemical Parameters

Some of the general indicator parameters used in conventional testing for organic chemicals include tests for Total Organic Carbon (TOC), Chemical Oxygen Demand (COD), and Biochemical Oxygen Demand (BOD), and UV absorbance (e.g., at 254 nm). UV absorbance and TOC analyses can be performed in minutes (on-line systems are available), COD takes hours, and BOD takes days. UV absorbance is not a particularly sensitive test for detecting spill events, and TOC is more commonly used for early warning purposes (Grayman et al. 2001).

Pesticides

Pesticide contamination of source waters is often seasonal as it primarily results from non-point source rainfall runoff from agricultural areas during periods of high pesticide application. Detailed quantification of pesticides (herbicides and insecticides) usually involves chromatographic separation and analysis, often coupled with mass spectrometry. Though applicable to on-line analysis, this technology is very expensive, and other less expensive (though less accurate) methods are available for frequent monitoring applications, including a relatively inexpensive immunoassay technique. For example, the immunoassay method for atrazine can be performed in approximately 40 minutes, is fairly inexpensive, and test kits are available from a variety of manufacturers. Of the many herbicides and insecticides that can pose risk to water supplies, atrazine (2-chloro-4-ethylamino-6-isopropylamino-*s*-triazine) is the one most commonly encountered in the United States, and is the primary herbicide used for control of annual broadleaf and grass weeds in corn and sorghum crops in the United States.

EARLY WARNING MONITORING

Rapid response to spills and other sudden changes in source water quality is necessary to protect water consumers from potentially harmful contaminants, determine appropriate changes in treatment strategy, and ensure compliance with environmental regulations. Early warning monitoring systems can provide a means to monitor short-term and long-term changes in water quality, and are an essential component of any rapid response program.

The primary goals of early warning monitoring systems are to allow operators to prepare for, detect, and respond to water quality changes such that public health is continually protected. Early warning and real-time monitoring can be incorporated as part of a utility's strategy for dealing with AFO contaminant releases, as well as releases from other contaminant sources.

Perhaps the greatest potential for significant sudden change in surface water quality caused by a CAFO is the breach of a large liquid waste storage facility.

Rapid or continuous real-time early warning monitoring can be used to detect accidental releases of certain contaminants as well as contaminant sources that have not been identified or properly controlled. How utilities respond to these events is important in terms of safeguarding the health of their customers and the integrity of certain treatment processes. A source water protection program can reduce the probability of spills occurring, while early warning notification of a contamination event allows utilities to respond appropriately to reduce potential adverse impacts, including possibly shutting intakes or adjusting their location, switching to an alternate source, adjusting or adding treatment processes, and increased monitoring.

Early warning systems are designed to provide utility personnel with timely information about water quality events that may otherwise pass through treatment (perhaps undetected) or result in a major process upset, thus allowing the operators to improve treatment reliability and effectively treat episodic events. Optimally, early warning systems provide this information continuously on a real-time basis.

The scope of an early warning monitoring program will depend on site-specific characteristics. Certain components, however, are generic to all good early warning systems, and include:

- Detection (a mechanism to detect pollution events)
- Characterization (a means to characterize the event)
- Communication (of data to utility personnel and other decision makers, of response actions to the public, etc.)
- Response (actions that minimize the potential impact of the contamination event)

For more sophisticated systems, organizational, institutional and communication infrastructures also need to be developed for managing and disseminating the information. Each of these components is discussed in various references (e.g., Grayman et al. 2004; Gullick et al. 2003; Gullick 2003; Grayman, Deininger, and Males 2001; and references therein).

Contaminant event detection mechanisms may include physical, chemical, microbiological and radioactive analyses, biomonitoring (bioalarms), physical observations, and reporting by other parties. The analytical monitoring methods seek to identify and quantify either a specific water quality parameter, or a surrogate parameter selected to provide a conservative indication of the presence of a more harmful contaminant. They include simple physical analyses such as turbidity, conductivity, and odor tests, relatively simple analytical tests for chemicals (e.g., specific ion electrodes for pH, nitrate and other ions; fluorescence for oils; immunoassays for herbicides), more advanced monitoring for chemicals (e.g., chromatography and mass spectrometry methods), and microbiological methods. However, the identification and quantification methods for many microbial contaminants are not very rapid, and real-time monitoring devices are not currently available for many of these.

Bioalarm systems use living organisms (e.g., bacteria, algae, zooplankton, bivalves, and fish) as sensors for extreme changes in water quality, but do not indicate the specific cause. Physical observations can detect visually-observable changes in water quality (e.g., oil sheens), and can include visual inspection or video surveillance of intake water or potential contaminant sources. Reporting by other parties can include notification of spill events by local, state or federal agencies, by the dischargers themselves, or by the observing public.

Some of the more common parameters examined in contaminant early warning systems for drinking water supplies include the following:

- On-line turbidity analysis as an indication of increases in solids concentrations and potential increases in microbial concentrations
- Multiparameter on-line probes for pH, dissolved oxygen, nitrate, ammonia, chlorophyll or other select chemicals depending on site-specific risks and needs
- On-line analysis of oxidant demand, or alternatively of total organic carbon
- Frequent batch immunoassay measurements of pesticide concentrations during their season of use

Several references are available that provide detail about rapid or on-line monitoring techniques for the water industry (e.g., Frey and Sullivan 2004; Gullick et al. 2003; Gullick 2003; Hargesheimer, Conio, and Popovicova 2002; Grayman, Deininger, and Males 2001; Pollack et al. 1999). A thorough review of the types of monitors used in numerous early warning monitoring systems and networks throughout the world is provided by Grayman, Deininger, and Males (2001). Information on water quality monitoring is also available from the intergovernmental National Water Quality Monitoring Council (www.nwqmc.org), the USEPA (www.epa.gov/OWOW/monitoring), and links referenced therein. Numerous publications are also available that describe the design of early warning monitoring systems (e.g., Gullick et al. 2003, Grayman et al. 2004, and references therein).

MICROBIAL AND CHEMICAL CONTAMINANT SOURCE TRACKING

Contaminant source tracking can be used for identifying the specific source of certain chemical and microbial contaminants, and for eliminating potential sources from consideration. Chemical or microbial 'fingerprints' can be analyzed and compared for identifying sources of those contaminants.

For chemical contaminants, various environmental sources have distinctive elemental isotope ratios for elements, and environmental processes can have impact on isotopic levels and ratios (Kendall 2004, Lindsey and Koch 2004, Moore et al. 2006). One example use of isotopic analysis in agricultural settings is tracing the source of nitrate found in groundwater through analysis of isotopes of nitrogen and perhaps oxygen. For example, soil nitrate may be compared to nitrate from an animal waste source through analyses of N^{15} (e.g., Moore et al. 2006). There is basically none of the N^{15} isotope in commercial fertilizer, but it is present in animal and human waste. However, this method does not usually work to distinguish septic waste from animal manure or to differentiate among manure from different kinds of farm animals, and in these cases a multi-isotope or multi-tracer approach would be better. N^{14} denitrifies faster than N^{15}, so in groundwater systems one would need to distinguish the denitrification process from the source.

Some options that are used less frequently are isotopic analyses of nitrogen (N^{15}) in dissolved organic nitrogen and ammonia, and oxygen (O^{18}) in phosphate or dissolved oxygen. USEPA, USDA, USGS, and many universities have isotope laboratories that can provide isotopic analytical services and possibly collaborations (Kendall 2004). More research and related work is recommended to make isotope methods more readily applicable (USEPA 2004b).

"Microbial source tracking" (MST) is a rapidly expanding science designed to identify human or animal sources of fecal pollution in the environment. These methods can evaluate *E. coli, Giardia*, *Cryptosporidium*, and viruses, and identify whether the source is from human, cattle, swine, bird, or other origin. Fecal source tracking using DNA "fingerprinting" can be used by water utilities to identify whether or not a particular potential fecal source in their watershed, such as an AFO or CAFO, is an actual contributor to wastes identified in the source water. Microbial source tracking analyses can cost a few hundred dollars each, so the sampling program design should be carefully planned and targeted.

Currently, the standard for most microbiological water quality monitoring utilizes the identification and enumeration of traditional and alternative microbial fecal indicator organisms. These indicators include total coliforms, fecal coliforms, *E. coli*, enterococci, and several other microorganisms that are associated with the human and animal intestinal tract. The premise of this approach is that these intestinal bacteria may also indicate the presence of other human and animal fecal pathogens that reside in the intestinal tract and may pose a risk to human health. The caveat to the indicator approach is that recent research has shown these indicators have limited utility for this purpose, citing differences in ecology, prevalence, resistance to stress, and the existence of environmental reservoirs in areas with no known fecal input. The result is a water quality monitoring system that may fail to accurately predict potential health impacts and offers no tools or information regarding remedial actions (Scott et al. 2002).

Source tracking methodologies take the indicator approach one step further by identifying the source or sources of these organisms. By identifying specific sources, better conclusions can be drawn regarding the possible presence of other pathogenic bacteria, viruses, and parasites. In addition, once a source is implicated, more appropriate steps can be taken to remedy the problem. The success (or failure) of a source-tracking project lies in the types of methods employed, the specific approaches taken, and the types of analytical tools used to interpret the results.

Numerous references are available that summarize the capabilities and state-of-the-science of microbial source tracking, including the following:

- USEPA Microbial Source Tracking Guide Document (USEPA 2005b) www.ces.purdue.edu/waterquality/resources/MSTGuide.pdf or www.calcoast.org/news/MSTGuide.pdf
- Scott et al., 2002 (http://aem.asm.org/cgi/content/full/68/12/5796)
- Bitton, 2005 (www.florida-stormwater.org/pdfs/FSAMicrobialSourceTrackingReport.pdf)
- USGS - Selected Bibliography for Topics Related to Microbial Source Tracking: http://water.usgs.gov/owq/MST_bibliography.html

Long et al. (2002) investigated alternative bacteriologically-based microbial indicator testing methods for human and grazing animal fecal contamination of drinking water sources. Monitoring *for Rhodococcus coprophilus* in tributary samples in close proximity to grazing animal activities can be used to indicate whether fecal matter is reaching the receiving water, and sorbitol-fermenting *Bifidobacteria* can be used as a human and wastewater source delineation tool to supplement coliform testing. Since these tests require a modest amount of labor, it is recommended that their use be limited to long-term watershed management and source water protection planning programs (Long et al. 2002).

Case studies using microbial source tracking for identification of fecal waste sources in the environment, including from animal feeding operations, are provided in the USEPA Microbial Source Tracking Guide Document (USEPA 2005b). To further help guide water utilities to best use microbial source tracking to their advantage, the process by which a source tracking study should be developed is outlined below:

- Specific knowledge of the watershed and potential fecal inputs is a critical component of a successful source tracking study. By concentrating only on potential inputs, the methods employed can be specifically targeted to identify these sources.

- All methods should be employed in a tiered fashion and results should be interpreted sequentially so as to implicate or eliminate potential sources one at a time.
- The sampling approach should be well planned and should be conducted during both high flow (wet) and low flow (dry) events. Targeted sampling should be performed, where samples are taken both upstream and downstream of potential fecal inputs. If bacterial counts are low upstream and high downstream (and source tracking data indicate a strong correlation with the suspected source(s)), then this is a strong indication that the source has been identified. Conversely, this approach can also eliminate suspected sources and redirect attention to other potential impact sites.
- End-users are encouraged to adopt a "toolbox" approach to any source tracking study. By using multiple methods that target and identify different source indicators, more comprehensive conclusions can be drawn. The most important aspect of the toolbox approach is that individual results can be confirmed and substantiated. Conclusions regarding potential inputs should never be drawn due to the results of a single test on a single water sample. On the contrary, multiple positive results using a variety of methods are a strong indicator that the source(s) of the pollution have been identified.
- If budgetary constraints are present, end users are encouraged to conduct their source tracking studies in a piecemeal fashion. It is better to target one or two potential sources with several tests, rather than to try to pinpoint all the potential sources with only one test per suspected source. Most source tracking projects fail to render meaningful results because too many sources of fecal pollution are targeted simultaneously. By limiting the number of potential targets, the end user can work by a process of elimination towards the likeliest sources of contamination.

UTILITY RESPONSES TO SUDDEN CHANGES IN WATER QUALITY

As discussed above, early warning monitoring systems can be used to detect water quality changes, and thus utilities can then take response actions such that public health and treatment systems are protected. For example, early warning systems can be incorporated into operational practices at the water plant to provide information to help decide on appropriate response actions to sudden changes in water quality. Potential responses may include one or more of the following:

- Increased monitoring and surveillance
- Spill cleanup
- Installation of protective booms or surface sprays near intakes
- Closing surface water intakes to let contaminants pass
- Repositioning intake position to a lesser or unimpacted location, if available
- Use of an alternate source of water supply
- Filling storage tanks prior to contamination reaching intake

- Enhanced water treatment (e.g., modify treatment process conditions such as using more coagulant for removal of particles, using additional disinfectant, and using greater amounts of powdered activated carbon)
- Shutting down water treatment plant
- Notification to public health agencies and the public, including public caution recommendations (e.g., for immuno-compromised) or issuing a boil water advisory
- Follow-up epidemiological studies
- No action

It is important to determine potential responses and response thresholds in advance of any potential contamination events. The principles of strong emergency contingency planning perhaps apply more to this component of an early warning monitoring system than to any other. Appropriate water supplier responses depend on the type and potential extent of contamination, projected risks (e.g., resulting effects on public health or treatment process efficiency), and characteristics of the affected population.

Water utilities with CAFOs in their watershed that store large amounts of liquid animal waste should have a contingency plan in place. In addition, they should make an effort to be included on all spill-related call lists used by state or local spill response agencies. Good communication with local emergency spill response organizations and state environmental agencies will help to ensure utilities are informed of spill events and provided information vital to a coordinated response. Riverbank filtration, pre-sedimentation, off-stream storage, and other pre-treatment structures can provide utilities with more options when dealing with these events, particularly large, short term, accidental spills and releases.

PRE-SEDIMENTATION BASINS AND OFF-STREAM STORAGE

Off-stream raw water storage and pre-sedimentation are similar in many respects because both act as additional barriers to passage of *Cryptosporidium* and other contaminants, and both provide general benefits to subsequent treatment processes by equalizing the raw water source, including dampening of both hydraulic and water quality spikes. However, off-stream storage generally refers to structures such as lakes or reservoirs with storage times on the order of weeks, versus pre-sedimentation that refers to tanks and retention basins with storage times on the order of hours or days. Another difference is that metal salts, organic polymers, or recycled softening sludge are commonly added to enhance physical separation within the pre-sedimentation basin, though some facilities do not add coagulant on a regular basis.

The LT2ESWTR allows *Cryptosporidium* removal credit for pre-sedimentation under certain conditions. With pre-sedimentation, utilities may either sample their source water after the pre-sedimentation basin and use that for determining bin assignment, or else use the pre-sedimentation system for *Cryptosporidium* removal credit. If off-stream storage is used, then sampling for bin assignment is done downstream of that storage, but the storage can not be used for removal credit under LT2ESWTR.

Pre-sedimentation is also similar to conventional clarification, except that pre-sedimentation basins are only intended to meet pretreatment objectives and, consequently, are often operated at higher overflow rates than conventional clarification. When coagulant addition is occurring and the treatment process is being actively managed to achieve clarification, solids removal achieved in pre-sedimentation reduces the solids load on the primary clarification process in the water treatment plant. The water treatment plant can operate the two processes in a complementary fashion so that the overall treatment process is less dependent on the removal of solids during filtration.

Pre-sedimentation is a useful tool in this process because it can help manage the initial particulate load reaching the primary treatment process. Pre-sedimentation not only initiates the solids removal process but it also helps to equalize water quality reaching the primary treatment process. The equalization, for example, can help dampen impacts of short periods of high contaminant concentrations (i.e., a chemical spill event). Such an approach is used by the City of Philadelphia Water Department, and others (AWWA 2004).

Data regarding the removal of *Cryptosporidium* and total aerobic spores (including *Bacillus subtilis*) during operation of full-scale pre-sedimentation basins have been collected independently by personnel from four US utilities as cited in AWWA (2004). *Cryptosporidium* oocysts were rarely detected in raw water at these locations and, consequently, it was not possible to calculate oocyst removal through these processes. Spores were detected more readily in raw water than were *Cryptosporidium*, making it possible to calculate log removal of spores at these utilities. At these facilities, spore removal typically ranged from about 0.5 to 1.0 log for pre-sedimentation basins operated between 0.3 and 2.8 gpm/ft^2. All of these basins either included addition of an organic polymer, a metal salt coagulant (e.g., ferric sulfate), or recycled lime softening solids.

One of the benefits of off-stream storage is that an off-stream storage reservoir can be isolated from the raw water source when needed, for example during the first flush after a storm or during a chemical or other spill. The treatment plant can operate using the reserve storage in the reservoir until the contamination in the raw water source passes the intake. LeChevallier et al. (1998) in a study on the Delaware River noted that peak levels of microbial contaminants following storm events were associated with peak turbidity spikes. By discontinuing pumping from source water during this period, the authors noted that *Cryptosporidium* spikes of 12 to 16 times baseline levels could be avoided. This illustrates that off-stream storage coupled with proper intake management practices can allow operations personnel to discontinue raw water pumping to avoid exposure to high risk events, such as first flush following a storm.

Oskam (1995) identified equalization of raw water quality peaks (chemical and microbial) and sedimentation of microorganisms and particles as some of the positive benefits of off-stream storage. The author also identified algal growth and contamination from runoff, if uncontrolled, and contamination from roosting birds as potential negative impacts on water quality associated with poorly maintained off-stream raw water storage structures. The factors identified by the author that can enhance the positive benefits of off-stream storage include the following:

- Higher degree of compartmentalization (three or four smaller reservoirs, or segments of reservoir systems, are better than one large uncompartmentalized structure)
- Greater hydraulic residence time
- Hydraulic characteristics of reservoir (shape) – this item is related to the first item
- Higher quality raw water entering reservoir

Several studies have investigated the reduction of *Giardia*, *Cryptosporidium*, and other microbial indicators during off-stream reservoir storage in the Netherlands (van Breemen et al. 1998, Ketelaars et al. 1995, Oskam 1995, Medema et al. 1998). These data indicate an average of 1.7-log *Cryptosporidium* removal during 5 or more months of off-stream storage for one system, and about 1.3 log after 70 days storage in another reservoir. Bertolucci et al. (1998) report data from Turin, Italy demonstrating mean *Cryptosporidium* removals of about 1 log after 18 days of storage.

LeChevallier and Au (2001) have summarized many of the literature sources currently available that address the removal or reduction in viability of *Cryptosporidium* and other microbial indicators during off-stream storage. Appendix 2 of the AWWA formal comments to USEPA on the LT2ESWTR cited earlier (AWWA 2004) includes literature references for studies investigating microbial removal and inactivation in off-stream raw water storage, especially focusing on *Cryptosporidium*.

RIVERBANK FILTRATION

Another pre-treatment strategy that is particularly well suited to situations where AFO wastes and other contaminants are possibly present in surface waters is to replace traditional surface water intakes with wells and other subsurface collection systems located adjacent to the surface water source. These systems are commonly referred to as riverbank filtration (RBF), although these systems can be located at the riverbank, lake bank, river bottom, or lake bottom. RBF essentially uses the subsurface media between the source and groundwater well to remove a variety of contaminants. RBF systems are robust treatment solutions that under appropriate circumstances (several dozen feet of unconsolidated, unfractured porous media) can reliably remove microbial pathogens, organic contaminants (including taste and odor causing substances and disinfection byproduct (DBP) precursors), hormones (e.g., Kolodziej, Harter, and Sedlak 2004), endocrine disruptors, turbidity, and other substances. RBF dampens or eliminates contaminant spikes (e.g., spills or other large accidental releases). RBF can also decrease impacts of diffuse pollution over space, by spreading over time and decreasing contaminant concentration.

RBF is particularly well suited for *Cryptosporidium* removal (Gollnitz et al. 2005, EE&T 2004). USEPA allows direct *Cryptosporidium* removal credit in the LT2ESWTR for new RBF systems. In this instance, *Cryptosporidium* bin assignment samples are collected from the surface water source and utilities that install RBF are allowed 0.5 or 1.0 automatic credits if they meet certain requirements, or higher credits through demonstration. Existing RBF systems are required to collect *Cryptosporidium* bin assignment samples from the well, consequently resulting in an indirect credit because the RBF well will likely result in a lower bin assignment than would samples collected prior to RBF. RBF systems typically are followed by conventional surface water treatment processes, including granular rapid-rate filtration. However, USEPA can allow some RBF systems that achieve suitable performance to convey water direct to consumers without subsequent treatment, other than disinfection.

Many utilities will be looking at RBF to reduce taste and odor complaints, provide protection against microbial contaminants, and reduce DBP precursors. RBF is also helpful in terms of security for source water quality, as the source would be more difficult to intentionally contaminate. Facilities with proper hydrogeologic settings for these installations can use RBF to protect themselves from impacts of a number of chronic and acute contaminant releases. Furthermore, RBF is becoming more accepted by the regulatory community as a potential treatment method for compliance with the USEPA Surface Water Treatment Rules (Gollnitz et al. 2005). Select successful case studies of utilities using RBF in agricultural watersheds are discussed in Chapter 7.

Studies reported by Wang, Hubbs, and Song (2002) indicate that organic compounds, turbidity, total coliforms, HPC (heterotrophic plate count – also called standard plate count), aerobic spores, algae, diatoms, Giardia, *Cryptosporidium,* insects, and other contaminants were found in Ohio River water but were either not found or were found at drastically reduced levels in adjacent RBF wells near Louisville, Kentucky. The collector wells for this system are located approximately 100 ft from the edge of the river, with collectors extending underneath the river at a depth about 50 ft below the river bottom. The water system uses a combination of RBF wells and direct intake from the river as source water for two conventional (coagulation, clarification, filtration) treatment plants.

Results from these studies indicate that the poorest removal occurred in the first few months after pumping began, indicating some acclimation time may be necessary for the microbiological removal processes occurring in the subsurface media between the river and the collectors. Typically, removal was also greater at greater distance from the river, though the greatest removal appeared to take place at the river interface and in the subsequent two or so feet adjacent to the river (Wang, Hubbs, and Song 2002).

Raw water turbidity during the study period ranged from 3 to 599 ntu, with mean and median of 49 and 17 ntu, respectively. Even during the 6 month or so acclimation period, turbidity in the wells was low, never exceeding 0.7 ntu. Furthermore, after the acclimation period the well turbidity levels were stable at about 0.15 ntu or less (95 percent $\leq$0.13 ntu). Turbidity spikes in the river caused some response in the wells during the acclimation period, but the wells became less and less influenced by river turbidity levels as the RBF system (i.e., subsurface media between river and well) matured. Similar findings were observed for coliforms (9 to 162,500 cfu/100 mL in river versus few detections, and none typically above 1 cfu/100 mL in wells), HPC (10 to 8,820 cfu/mL in river versus typically <10 cfu/mL in well), aerobic spores (mean raw 230,000 spores/L versus typically <1 spore/L in well), algae (>100,000 algae/L in raw versus typically no detections in well), and TOC (mean 2.9 mg/L in river, mean 2.1 mg/L nine feet from river, to 1.6 mg/L in well). For the organics, it was estimated that about 60 percent of the removal could be attributed to biodegradation and the remainder to physical removal processes (e.g., sorption and filtration). Removal of DBP precursors followed similar trends to TOC removal, with about 60 percent measured reduction in haloacetic acid (HAA) precursors and approximately 40 percent removal of trihalomethane (THM) and total organic halogen (TOX) precursors (Wang, Hubbs, and Song 2002).

TREATMENT AT WATER PLANT

For hydrogeologic settings where RBF is not suitable, other alternatives are available, including: ultraviolet (UV) disinfection, membranes, alternative oxidants, and other specialized treatment. These alternatives are more complex to operate than RBF, but are also becoming more common. For example, recent improvements in UV and membrane systems have made the systems more reliable, simpler to operate and maintain, and more cost-effective to operate. Regulatory pressure requiring compliance from the LT2ESWTR and the Stage 2 Disinfectant Byproduct Rule (DBPR), for example, is influencing utilities to look more closely at these technologies. Situations where chronic or acute releases of contaminants from AFOs are a potential problem may provide a further incentive or justification for employing these technologies.

Several utilities have installed activated carbon adsorption, UV disinfection, advanced oxidation, or membrane filtration at least in part for treatment of contaminants originating from CAFO operations, including five of the 30 water utility respondents to a survey for this research project that had CAFOs located in their source water area (the survey results are described in Chapter 7).

USE OF WATER TREATMENT RESIDUALS TO STABILIZE PHOSPHORUS IN ANIMAL WASTES

Stabilization of nutrients by alum (aluminum sulfate) is a relatively new area of research (USEPA 2004b), and an area where a water utility can use their own treatment plant's residuals to help in source water protection from agricultural operations. Beneficial use through application of alum and ferric hydroxide water treatment residuals to agricultural land can provide an inexpensive disposal option for utilities and protect source water quality by binding to and reducing the transport of phosphorus (DeWolfe 2003, DeWolfe 2006). The use of alum on poultry litter can help stabilize phosphorus to prevent its runoff to surface waters and leaching to ground waters.

Alum can be added to dry litter in poultry houses, typically between shifts of birds in the houses. The alum binds the phosphorus in the litter and makes it less soluble. Up to 90 percent less phosphorus may be solubilized in runoff from fields with alum-treated litter versus untreated litter (Walker 2000). Alum also reduces ammonia gas emissions from the litter, which in turn improves bird health, increases weight gains and reduces mortality, thus saving the producer money. In fact, savings from these benefits may more than offset the cost of the alum (Walker 2000).

DeWolfe (2006) evaluated the ability of different types of water treatment residuals to reduce loss of phosphorus from agricultural land, developed recommendations for utilities regarding use of their residuals, and demonstrated land application of residuals as a best management practice. The research found that each water treatment residual (WTR) has unique phosphorus binding potential based on the coagulant used, age of the WTR, and form of the WTR used (wet or dry). Blending WTRs with a high-phosphorus soil demonstrated reductions in both soil test phosphorus and soluble phosphorus. Simulated rainfall testing confirmed the ability of WTRs to reduce phosphorus in runoff for soils that are high in phosphorus and have also recently received organic material (e.g., poultry manure). The WTRs applied as liquid reduced phosphorus slightly better than when applied as dry material. WTRs blended with biosolids and poultry manure revealed substantial reductions in both soil test phosphorus and soluble phosphorus, and that the reduction was directly related to the amount of WTRs used. An ongoing related AwwaRF-sponsored research project on the "Effect of Long-Term Water Treatment Residuals-Biosolids Co-Applications on Native Soil Phosphorus" (AwwaRF Project #2995) is investigating both the long-lasting effects of a single water treatment residuals-biosolids co-application and the short-term impacts of a repeated application on soil phosphorus dynamics.

REGULATORY ENFORCEMENT AND LEGAL ACTIONS

Water utilities concerned about CAFO impacts on source water quality may want to confirm that the facilities are meeting the appropriate regulatory requirements and deadlines for permitting and implementation of permit and nutrient management plan requirements. Communication with the appropriate regulatory agency that oversees CAFO NPDES permitting can help to identify sites with substandard operating procedures or facilities. Pressure can be put on the regulatory agency to take action to ensure these facilities implement measures to be in compliance with the relevant requirements.

As a last resort, some utilities have felt it necessary to take legal action against CAFO interests. Good, sound scientific assessments of source area pollutant loads and impacts are essential for presenting a valid case to force producers to implement pollution prevention measures. As an example, the City of Tulsa, Oklahoma spent several years trying to work with local poultry producers in the Eucha/Spavinaw watershed to reduce phosphorus loads that caused eutrophication and taste and odor issues in the source water. Attempts at stakeholder cooperation were not successful in meeting the City's primary goals, and they eventually filed a legal challenge against the poultry integrators in the watershed. Supported by good scientific assessments of the phosphorus issues in the watershed, Tulsa was successful in getting a substantial reduction required in the amount of chicken litter spread on land in the watershed, as well as a reduction in phosphorus discharges at a municipal wastewater plant whose flow is primarily from a large chicken processing facility.

CHAPTER 7
CASE STUDIES

This chapter includes information on various case studies selected to represent a variety of water utility approaches and CAFO types. The examples are provided to encourage water utilities to consider alternative approaches for their particular situation. Three sections of case studies are provided:

1. A series of summaries of case studies for agricultural watersheds with AFOs and CAFOs that are described on USEPA websites
2. Experiences described by US water utilities that participated in an informational survey conducted during the project
3. Select in-depth case studies analyzed as part of this research project, including:

   - Dairy cattle operations (Syracuse, New York)
   - Beef cattle operations (Amarillo, Texas)
   - Poultry operations (Tulsa, Oklahoma)
   - Swine operations (Wilmington, North Carolina)
   - Dairy cattle operation impacts on groundwater (San Joaquin Valley, California)
   - Riverbank filtration (Lincoln, Nebraska; Cedar Rapids, Iowa; and Mankato, Minnesota)
   - Stakeholder cooperation (Philadelphia, Pennsylvania)

Site visits and interviews were conducted at AFO and CAFO facilities, water utilities, environmental regulatory agencies, Soil and Water Conservation Districts, and Cooperative Extension offices as part of the case studies in New York, Texas, Oklahoma, and North Carolina. A map showing the locations of each of these case studies is presented as Figure 7.1.

Figure 7.1 Locations of case studies

CASE STUDY SUMMARIES COMPILED BY USEPA

USEPA has provided descriptions of various successful watershed and source water protection case studies (USEPA 2005a, 2005e, 2000, 1999), including the following examples that include experience with agricultural issues and BMPs. The descriptions below focus on the water quality issues and BMPs implemented. Additional information is available on the referenced web sites, including partnering organizations and funding sources.

- Springfield, Missouri (USEPA 2005 www.epa.gov/safewater/protect/casesty/springfield.html) – well-known agricultural program, including demonstration BMP project sites for educating farmers.
- Upper and Lower Bad River Watersheds (central South Dakota) (USEPA 2000; http://water.usgs.gov/owq/cleanwater/success/bad.html) – includes erosion BMPs, monitoring and education programs; results suggest sedimentation reaching the Missouri River has been reduced by 30 percent, and measured success in one subwatershed showed a drop in sediment loading from 82.7 down to 10.2 tons per acre/foot of runoff over a period of five (5) years.
- North Fork of the Ninnescah River Watershed (south-central Kansas) (USEPA 2000; http://water.usgs.gov/owq/cleanwater/success/ninnescah.html) – includes numerous agricultural BMPs, as well as on-farm demonstrations, farm tours, educational workshops and farmer-to-farmer meetings. In addition to the watershed program obtaining traditional cost-share funding (e.g., USDA's

Environmental Quality Incentive Program), the City of Wichita contributes funding for BMPs so that in some cases farmers incur no expenses.

- Little Rabbit River Watershed (southwest Michigan) (USEPA 2000; http://water.usgs.gov/owq/cleanwater/success/rabbit.html) – efforts for sedimentation and nutrient control include a partnership that created a resource management system involving 16 different BMPs on over 17,000 acres.
- Skaneateles Lake Watershed (Syracuse, New York) (USEPA 2005; www.epa.gov/safewater/protect/casesty/skaneateles.html) – effective watershed agricultural program (a separate case study description prepared specifically for this research project is described below).
- Lower Truckee River (Pyramid Lake Indian Reservation, Nevada) (www.epa.gov/owow/nps/Success319/state/nv_pp.htm) – uncontrolled cattle grazing and poor agricultural practices resulted in the loss of instream and riparian habitat, degrading the water quality of the Lower Truckee River and Pyramid Lake. Livestock fencing, laser leveling of agricultural fields, and other agricultural BMPs have successfully reduced in-stream nitrogen levels and led to achievement of water quality standards in the river.
- Tar-Pamlico River Basin, North Carolina (www.epa.gov/owow/nps/Success319/state/nc_tar.htm) – row crops and AFOs in the basin, one of three main feeders to the US's second largest estuary (the Albemarle-Pamlico Sound) led to excessive nutrients in the estuary, eventually resulting in it being added to the state's 303(d) list for chlorophyll *a*. Reduced fertilizer use, riparian buffers, implementation of conservation tillage practices and other agricultural BMPs led to a 30 percent reduction in nitrogen loads and allowed one section of the estuary to be removed from the 303(d) list.
- Neuse River (North Carolina) (www.epa.gov/owow/nps/Success319/state/nc_neu.htm) – the Neuse River is one of three main feeders to the Albemarle-Pamlico Sound. High nitrogen levels in runoff from crops, pasture land, and AFOs led to frequent algal blooms, hypoxic conditions, and fish kills in the river. Implementation by the agricultural community of various BMPs, including buffers, contour planting, no-till planting, and creek fencing, helped contribute to a 42-percent decrease in nitrogen loading to the estuary (the TMDL goal was a 30-percent reduction). Combined with additional point source reductions, these measures helped result in a 27-percent reduction in nitrogen levels in the Neuse River just upstream of the estuary.
- Blue Spring Creek (Coffee County, Tennessee) (http://www.epa.gov/owow/nps/Success319/state/tn_blue.htm) – runoff from livestock operations and unrestricted grazing contributed high levels of sediment and nutrients to Blue Spring Creek. An education program and implementation of agricultural BMPs (including livestock fencing, cattle watering facilities, and waste management systems) helped to reduce the water quality problems, allowing the creek to be removed from the state's 303(d) list.
- Cane Creek (McMinn County, Tennessee) (http://www.epa.gov/owow/nps/Success319/state/tn_cane.htm) – Cane Creek was contaminated with pathogens due to urban (runoff/storm sewers) and agricultural (pasture grazing) sources. Pathogen inputs were reduced by stabilizing erosion-prone areas near AFOs and relocating the City of Etowah's stormwater discharge point. As a result, the creek was removed from the state's 303(d) list.

- Little Shoal Creek (south-central Tennessee) (http://www.epa.gov/owow/nps/Success319/state/tn_little.htm) – Little Shoal Creek was impaired from high levels of silt coming from pasture grazing and other agricultural activities. Various pasture management practices were implemented to reduce erosion and pollutant transport to the creek, allowing the creek to be removed from the state's 303(d) list.
- North Fork of the South Branch of the Potomac River (northeastern West Virginia) (www.epa.gov/owow/nps/Success319/state/wv_nor.htm) – this scenic trout stream was impaired by high levels of fecal coliform bacteria, primarily a result of runoff from beef and poultry farms. Over 85 percent of the farmers in the watershed worked to construct various BMPs at the farms, including constructing animal waste storage facilities, establishing riparian buffers, and others. The steam is no longer listed as impaired.
- Bass Lake (Wisconsin) (www.epa.gov/owow/nps/Success319/state/wi_bass.htm) – livestock operations and other agricultural activities were determined to be the cause of high nutrient levels in Bass Lake and resulting fish kills. The Marinette County Land and Water Conservation District led an effort to reduce phosphorus levels in the lake through a combination of installing barnyard control practices and in-lake treatment techniques. TMDL targets were achieved by reducing the average phosphorus concentration from 490 μg/L down to 10 μg/L, and the lake will be removed from the state's 303(d) list during the next listing cycle.
- Cascade Reservoir (central Idaho) (http://www.epa.gov/owow/nps/Success319/state/id.htm) – Cascade Reservoir was adversely affected by high levels of phosphorus resulting from a variety of nonpoint sources (84 percent of the load) such as agricultural activities and forestry operations, point sources (10 percent of the load) such as wastewater treatment plant and fish hatchery discharges, and poorly functioning septic tanks (6 percent of the load). Implementation of grazing BMPs and upgrading forest roads (and other projects) resulted in a 57-percent reduction of phosphorus levels in the reservoir. Agricultural BMPs implemented included grazing and irrigation management through livestock exclusion, fencing, tree and shrub planting, and wildlife habitat management.
- Nooksack River (Lummi Indian Nation, western Washington) (www.epa.gov/owow/nps/Success319/state/wa_nook.htm) – high levels of fecal coliform bacteria in the river from dairy and livestock operations, municipal wastewater treatment plants, and poorly operating septic systems contributed to high levels of bacteria in Portage Bay, forcing the closure of shellfish harvests. A dairy inspection program was implemented, and the requirement for dairy nutrient management plants resulted in a 63-percent reduction in fecal coliform bacteria in the bay and the reopening of some of the shellfish beds.
- Sauk River Chain of Lakes (central Minnesota) (www.epa.gov /owow/nps/Success319/state/mn.htm) – the Sauk River Chain of Lakes is an interconnected systems of 14 bay-like lakes fed by the Sauk River. The waters are impaired by high levels of phosphorus and suspended solids due to row cropping and livestock operations, as well as discharges from private and public wastewater treatment systems. Implementation of agricultural BMPs and upgrades to septic systems and municipal wastewater treatment systems in the watershed reduces phosphorus loading by 48 percent. Agricultural improvements

included the installation or improvement of 50-agricultural waste storage facilities, development of more than 40 Manure Management Plans, installation of feedlot filter strips and retention basins, and enrollment of more than 5,000 acres into the USDA Farm Service Agency (FSA) Conservation Reserve Program. In addition, vegetative buffer strips were planted and shoreline areas restored to reduce erosion along riparian areas. In the early 1990s the Sauk River watershed was the first major watershed in Minnesota to implement a watershed-wide phosphorus discharge limit of 1.0 mg/L for wastewater treatment facilities. The requirement contributed substantially to improvements in the Sauk River and allowed water quality improvements from projects funded by the Clean Water Act Section 319 program to become more evident.

- Middle Fork Holston River (southwestern Virginia) (www.epa.gov/owow/nps/Success319/state/va.htm) – agricultural and residential activities contributed to the river becoming impaired by high levels of sediment and fecal coliform bacteria. Improvement measures focused on reducing discharges from private wastewater systems and excluding livestock from streams, and helped reduce bacteria levels in the creeks feeding the river. Farmers installed over 14 miles of stream fencing, implemented livestock grazing management systems (including watering sources and travel lanes), and improved 3,588 acres of pasture and 13 acres of riparian forest buffers along Hutton Creek. Approximately 70 percent of the agricultural producers in the watershed have been contacted about the goals of the watershed protection plan
- Dungeness River (Jamestown S'Klallam Indian Tribe, Washington) (www.epa.gov/owow/nps/Success319/state/wa_dungeness.htm) – high levels of bacteria have resulted in shellfish bed closures and the inclusion of the river and a tributary creek on the state's 303(d) list as impaired by fecal coliform contamination. Contributing sources included livestock management, failing septic systems, and pet wastes. The Clallam Conservation District (http://clallam.scc.wa.gov/) worked directly with farm operators to develop conservation plans for individual farms. Agricultural BMPs implemented included fencing, riparian restoration, pasture management, and manure storage. Piping of irrigation ditches (in areas deemed high priority) and eliminating tailwaters in the extensive irrigation network in the watershed were also important components of the program. Education and outreach efforts included newsletters, workshops, and presentations to help inform livestock operators about the impacts of livestock management on water quality
- Lower Yakima River (Washington) (www.epa.gov/owow/nps/Success319/state/wa_yakima.htm) – erosion from irrigated agricultural lands was blamed for high levels of suspended solids, turbidity and the pesticide DDT in the Lower Yakima River, thus causing the river to be included on the state's 303(d) list as impaired by these pollutants and a corresponding TMDL to be developed. A partnership of two irrigation districts in the watershed was formed, and they developed a Water Quality Policy (with support from local farmers and other landowners) that set specific on-farm turbidity targets. Over 200 site-specific plans have been prepared. The partnership worked closely with farmers who voluntarily converted a total of over 20,000 acres from water-intensive and erosive rill and furrow irrigation methods to more efficient sprinkler or drip systems. Other BMPs implemented to control erosion include the construction of settling ponds,

filter strips, and the use of polyacrylamide - a substance that binds to soil while allowing water infiltration. These measures have resulted in meeting the interim TMDL goals for turbidity at three of four primary irrigation water return drains, and significant progress was made at other monitoring sites as well

USEPA Region 4 also has a website with descriptions of several Clean Water Act Section 319 non-point source pollution projects related to nutrient contamination (www.epa.gov/region04/water/nps/projects/index.htm), many of which are from agricultural and animal feeding operations, including the following helpful examples:

- Duck River Watershed Project (poultry and beef cattle wastes) www.epa.gov/region04/water/nps/projects/al99-2.htm
- The Delta Farmers Advocating Resource Management (FARM) Project (primarily cropland BMPs) www.epa.gov/region04/water/nps/projects/ms00-1a.pdf
- Riparian Buffers and Controlled Drainage Evaluation (studied the effectiveness of vegetative buffers and controlled drainage for row crops, grazing dairy cattle, grazing beef cattle, and unfertilized pasture) http://www.epa.gov/region04/water/nps/projects/nc97-1.htm
- Water Quality Impact of Hog Lagoon and Turkey Litter in Loblolly Pine Stands in the Coastal Plain and Sandhills of South Carolina (forested land is an additional animal waste disposal option where agricultural land is becoming scarce; this project studied sound forest land application levels of hog lagoon effluent and turkey litter) www.epa.gov/region04/water/nps/projects/sc98-1b.htm

Another case study of interest is that of the Conesus Lake watershed in New York State. In support of implementation of the Conesus Lake Watershed Management Plan (http://co.livingston.state.ny.us/conesus.htm), a project is being conducted that represents an integrated approach of academic, extension and farming interests to conduct research at the watershed level that will test the impact of BMPs on mitigation of non-point sources of nutrient and soil loss. A description of the BMP research project is available at (www.envsci.brockport.edu/Conesus_Project/).

UTILITY SURVEY

A survey of US water utility experiences regarding AFOs and CAFOs was conducted in 2005. A copy of the survey is attached in Appendix B. The utility distribution list was developed starting with a list of AwwaRF subscribers. That list was thinned down in states where CAFOs were less likely to be common to make the number of surveys manageable. Then each state where CAFOs were more likely to be present was examined to see if there were additional cities that could be included. Several additions were made to the list from that analysis. The survey questions were developed by the project team, and then critiqued by the PAC before it was distributed.

The survey was sent to 182 water utilities, ranging from one to nine in all of the 50 states (except Alaska). 64 replies were received (35 percent of those sent); of those 30 have CAFOs in their watershed. Combined, these 30 water utility survey respondents produce over 10,000 mgd for over 16 million people in 18 different states. Figure 7.2 shows for each state how many utility surveys were received along with the number of responding utilities that have CAFOs in their watershed. Respondents with CAFOs in their watershed came from the following states:

Arkansas (1), Iowa (2), Idaho (1), Illinois (3), Minnesota (4), Missouri (1), New York (2), North Carolina (3), North Dakota (1), Nebraska (2), Ohio (1), Oklahoma (1), Pennsylvania (1), South Carolina (1), Texas (3), Utah (1), Vermont (1), and Wyoming (1).

The remainder of this discussion focuses on the responses from the 30 utilities with CAFOs in their watershed. Combined these respondents present a wide range of CAFOs in their watersheds, source water protection measures and goals, treatment responses, and degrees of communication with CAFOs.

The types of CAFOs in those 30 watersheds include swine, beef, dairy, chickens, turkeys, aquaculture, horse, buffalo, and elk. The primary CAFOs for these watersheds include 10 dairy, 9 beef, 5 swine, 3 chicken, and a few utilities did not know which were the dominant CAFOs. Pathogens and other microbial contamination were the current concern most frequently listed by these utilities (90 percent of the respondents), followed by nutrients and pesticides (48 percent), and less frequently pharmaceuticals, other organic contaminants, and metals (approximately one-third of the respondents listed each of these). Six utilities have direct evidence of CAFO impacts on their source water quality, twelve have no direct evidence but suspect an impact, and the rest have no known or suspected impact.

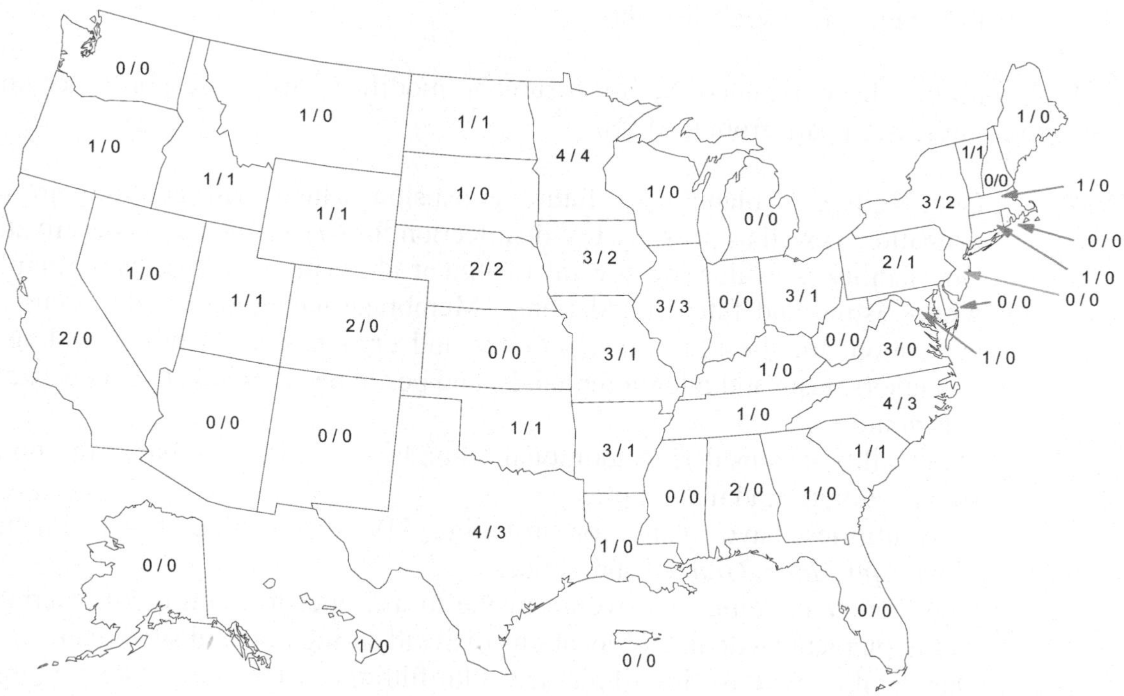

Figure 7.2 Number of survey responses and utilities with CAFOs in their watershed (# of utility responses / # of utility responses with CAFOs known in their watershed)

Six utilities have already instituted treatment modification in response to environmental water quality impacts perceived to be, at least in part, from CAFOs:

- One facility added UV disinfection treatment for *Cryptosporidium*, *Giardia*, and viruses
- One facility upgraded to ozonation in preparation to reduce DBP precursors and other natural organic matter (a future expansion is planned which will continue to utilize ozonation as well as possibly UV disinfection for *Cryptosporidium* deactivation)
- One facility installed ultrafiltration membranes for removing a variety of contaminants
- One utility uses additional algaecide (e.g., copper sulfate) in their reservoir for algae control, and also installed additional granular and powdered activated carbon due to increases in taste and odor compounds (geosmin and methyl iso borneol)
- Another facility also uses powdered activated carbon
- One facility changed their chlorination sequence in response to elevated haloacetic acid (HAA) levels that occurred after heavy rainfall and runoff following a five-year drought

Eight utilities have planned future treatment modifications in response to impacts perceived to be, at least in part, from CAFOs:

- One facility is planning a future expansion which will continue to utilize ozonation as well as possibly UV disinfection for *Cryptosporidium* deactivation
- One facility is evaluating several treatment alternatives to maintain compliance with existing and future regulations. Membrane technology is the focus of the evaluation, for the removal of nitrates and organic compounds including DBP precursors, taste and odor compounds, and pesticides of which CAFO wastes may contribute
- One facility is considering additional groundwater wells, riverbank filtration, off-stream storage, and membranes
- Two utilities have plans for installing UV disinfection for treatment for *Cryptosporidium*, *Giardia* and viruses
- One facility is going to move an intake to a deeper location below thermocline and implement switching of locations to avoid spring rain and snow melt
- One facility will be installing new ultrafiltration membrane filters to replace existing sand filters
- One facility is considering enhanced coagulation, microfiltration or UV disinfection

At least 19 of the 30 utilities have source water protection programs in place. Interestingly, ten of the respondents who said they had CAFOs in their watershed also noted that CAFOs were not identified in the Source Water Assessment Program (SWAP) report for their source water.

Cooperation with the local farmers is essential for the success of any agricultural source water protection program. Many of the participating water utilities present excellent examples of different approaches to take for communication and partnership with the local agricultural interests. Utility contacts with CAFOs include the following various examples:

- Participation in watershed coalitions that include state agencies, local watershed partners and farm operators
- Working with county conservation districts, Agricultural Extension agents, state regulatory agencies, state agricultural agencies, and USDA offices that have existing dialogues with CAFOs
- Direct work with a CAFO on a stream restoration project funded in part with a 319 grant, as part of a TMDL action item
- Participation in the Quad State Poultry Dialogue (Arkansas, Missouri, Kansas, Oklahoma)
- Meetings with the Soil and Water Conservation District (or the Natural Resources Conservation Service) and area land owners to discuss water quality issues, generally centered around nutrient loadings related to row crop activities
- In one watershed all dairies have agricultural management plans developed with assistance from local and state agricultural specialists. That same utility provides cost-share funds to agricultural operators who install qualifying agricultural Best Management Practices
- Another utility contracts with the local Soil and Water Conservation District to operate a Watershed Agricultural Program which develops Whole Farm Plans (including Nutrient Management Plans) for the participating farmers (mostly dairy), and the utility also pays 100 percent of the costs of all agricultural BMPs installed. Local farmers were directly involved with the water utility at each of the primary phases of development, implementation and review of this program. Annual meetings are held for the program to further engage local farmers
- One water utility says their Watershed Agricultural Program, which is funded by the utility and implemented by a Watershed Agricultural Council, has become a national model for addressing agricultural nonpoint source pollution. More than 95 percent of the watershed farms have joined the program, which develops Whole Farm Plans (WFPs) and recommends BMPs to reduce agricultural pollution and enhance the economic viability of participating farms
- One city established a stakeholder-based Comprehensive Watershed Management Team in 1997 to develop a regionally coordinated approach to address water quality degradation in their watershed. Participants were drawn from two states, and included poultry growers, poultry integrators, poultry litter haulers and applicators, concerned citizens, federal and multi-state agencies and universities. Three working groups were established to focus on technical issues and build a sound scientific and economic basis for addressing water quality degradation

The primary source water protection measure utilized is monitoring for priority contaminants and their surrogates. Other measures have included the following:

- Filter strip programs - riparian buffers
- USDA EQIP cost-share program to fence out cattle from ponds and creeks
- Pilot project using fencing to keep animals out of streams
- Public education campaigns and informational web sites

- One utility provides cost-share funds to agricultural operators who install qualifying agricultural BMPs. This program is coordinated with local and state agricultural agencies
- One utility pays for development of Whole Farm Plans, including NMPs, as well as 100 percent of the costs of installed BMPs at farms participating in their Watershed Agricultural Program. They also purchase conservation easements, conduct an educational program, and instituted a program to replace septic systems with composting toilets at private residences near the water supply lake
- Discussion with conservation districts, agricultural extension agents, and county and local planners and government officials. In some cases, farmers have been successfully dissuaded from establishing new CAFOs in particularly sensitive areas or without well-designed controls
- Working closely with the Farm Bureau to encourage CAFOs to voluntarily make improvement to their management of wastes and animal feeding practices
- One city funded and implemented programs including numerous scientific studies, an interactive Geographical Information System (available to the public), a USEPA-approved Quality Assurance Project Plan (QAPP), a poultry litter export hotline, a common data reference library, and the development of the state's first Watershed Restoration Action Strategy (WRAS). Numerous intensive studies were completed to address water quality problems by quantifying nutrient loadings, assessing impacts on water quality, and setting nutrient target values for source water area management
- One utility developed pollution prevention plans for 244 commercial farms (mostly livestock) in their source water area. Over 95 percent of the commercial farms in the source water area are participating, including all AFOs. Most are small farms, and about five of these farms have more than 300 animal units and only one has more than 1,000-animal units. The plans identify and recommend BMPs to address animal waste, nutrients, pathogens, soil erosion and pesticides. Since 1992, over 3,000 of the BMPs recommended have been implemented at a total cost (excluding planning and design) of over $22 million. Nutrient Management Plans were also provided to all livestock farms (including manure spreading and fertilization recommendations based upon soil samples), and 1,624 acres of riparian forest buffers have been or are in the process of being installed on farms
- Arkansas has recently enacted new poultry litter application regulations requiring waste management planning for poultry operations and application of litter in accordance with Arkansas phosphorus index

The 30 water utilities cited a variety of goals for future source water protection efforts related to potential CAFO impacts, including:

- Develop and implement source water protection plans and programs
- Develop an increased understanding of the status of CAFO operations and the impacts of CAFOs wastes on source water quality
- Increase monitoring, including monitoring over a long enough period of time to gain stronger understanding of CAFO wastes on the source water
- Increase communication and partnership with all parties involved

- Develop education and planning activities to help operators of AFOs and other farms understand the impacts of their operations on downstream potable water systems, and how they may lessen the potential impact from their facilities
- Develop strategies to keep wastes from entering waterways, and to prevent accidental discharge to surface waters
- Implement measures for reducing source water contamination, including encouraging the installation of BMPs
- Implement cost-sharing programs for animal waste and other agricultural BMPs
- Perform periodic follow-up of BMPs to ensure proper operation and maintenance, and to identify new environmental problems
- Provide Nutrient Management Plan updates on all participating livestock farms every three years
- Restore river and stream banks and keep AFOs away from natural channels
- Implement riparian buffers along all streams in the source water reservoir watershed
- Increase use of fencing to keep animals out of pond and creek areas
- Institute proactive approaches for remediation of problems that have the potential for occurrence
- Establish protective zones that prevent certain land uses around the water supply reservoir
- Upgrade classification of the source water to that of a drinking water source

DAIRY CATTLE OPERATIONS (SYRACUSE, NEW YORK)

City of Syracuse Water Supply

Skaneateles Lake has served as the primary water supply for the City of Syracuse and several outlying communities since 1894. The lake is approximately 15-miles long, with an average width of one half to three quarters of a mile, and is in a relatively small watershed of 59 square miles (38,000 acres). The watershed is comprised of portions of one village and seven towns within three counties (Onondaga, Cayuga, and Cortland). Approximately 48 percent of the land use in the watershed is agriculture, 40-percent forest, 5.4-percent private residential, and the rest is vacant or commercial. A large portion of the watershed is fairly steeply sloped.

Skaneateles Lake is oligotrophic, thus low on nitrogen, phosphorus and other nutrients. The primary concern to drinking water source quality is pathogens from agricultural practices (e.g., the SWAP for this water supply identified a high potential for protozoan contamination due to pasture land in the watershed). There are approximately 51 farms in the watershed, the majority dairy farms, and many of these are smaller family farms. There is one dairy CAFO located completely in the watershed, and parts of four other CAFOs within the watershed border. Additional information on the watershed and water supply is available at www.syracuse.ny.us/deptWater.asp.

The City of Syracuse Department of Water (www.syracuse.ny.us/deptWater.asp) operates with a filtration avoidance waiver for the Skaneateles Lake water supply from the New York State Department of Health (NYSDOH; www.nyhealth.gov). Treatment consists primarily of chorine disinfection for the average 43 mgd production that serves approximately 200,000 people.

The filtration avoidance waiver requires specific watershed rules and regulations as well as water quality protection programs. The filtration waiver will remain in effect as long as the

City maintains its watershed protection programs and the lake water continues to be of high quality. The filtration avoidance waiver also requires that the Skaneateles Lake intake be closed during high turbidity events to minimize the potential for occurrence of pathogens in the drinking water supplied by the City. During those events supply can be supplemented with treated Lake Ontario water purchased from the Onondaga County Water Authority.

City of Syracuse Source Water Protection Programs

To protect the quality of the lake and maintain the filtration avoidance waiver, the City of Syracuse Department of Water conducts and funds a variety of source water protection measures for the Skaneateles Lake watershed. The various activities are outlined on the Skaneateles Lake Watershed Website (http://www.cce.cornell.edu/onondaga) and in the watershed management plan (http://www.cce.cornell.edu/onondaga). The City employs a full-time watershed control coordinator to oversee the programs. The City has conducted a watershed protection program since shortly after it started using the lake as a water supply in 1894.

Source water protection activities are focused on the following potential pollutants and sources:

- Agricultural pathogens, especially *Cryptosporidium* and *Giardia*
- Sediments and turbidity, especially runoff issues associated with steep slopes and recent strong storm events
- Private wastewater systems
- Pesticide usage

The primary focus is on pathogens originating from agricultural practices and private wastewater systems.

Activities in the watershed that might affect water quality are controlled under New York State authority with a set of Watershed Rules and Regulations specifically enacted for Skaneateles Lake. The NYSDOH Watershed Rules and Regulations (NYCRR Title 10, Section 131.1; effective July 6, 2005; www.health.state.ny.us/nysdoh/phforum/nycrr10.htm - search for "Section 131.1 City of Syracuse") are required by the USEPA filtration avoidance waiver. The rules include requirements for private wastewater systems, protecting the lake from erosion from construction sites, solid waste management, agricultural practices, forestry, and other measures. New York State Environmental Conservation Law (ECL Article 17-1709) also prohibits the discharge of sanitary or industrial wastewater to the lake. The City of Syracuse Department of Water is responsible for making regular and thorough inspections in the watershed to enforce these rules and regulations.

There are five primary components to the City's watershed protection program:

- Watershed Agricultural Program (described below in more detail)
- Inspection Program: the watershed management plan includes regular inspections of numerous agricultural, commercial and residential properties in the watershed
- Public Education Program comprised of workshops, publications, websites, and demonstration projects
- Land Protection Program: purchases conservation easements on environmentally sensitive land (and includes a specific educational component for the property owners about environmental stewardship)

- Data Gathering and Management Program: Geographic Information System (GIS) mapping, identification of hydrologically sensitive areas and potential pollutant sources, water quality data, and ranking of parcels for inclusion in the Land Protection Program

Two full-time Department of Water inspectors patrol the lake and watershed looking for problem areas and violations of the NYSDOH Watershed Rules and Regulations and New York State Department of Environmental Conservation regulations, and then identify and target solutions (additional inspectors are hired in the summer). The inspectors walk properties, look under structures, inspect for sewage system problems (hundreds of dye tests are conducted on residential wastewater treatment systems each year), and investigate any construction activities.

The City also recently replaced private outhouses in the watershed with new composting toilets. The City historically provided "pail service" to the residents for cleaning outhouses. The compost material from the new toilets poses little or no health risk to workers and the environment, and comprises a far smaller volume than that of the outhouse waste.

According to the Watershed Rules and Regulations, animal waste storage and disposal in the Skaneateles Lake watershed must be performed either (a) in accordance with the publication "Agricultural Management Practices Catalogue for Nonpoint Source Pollution Prevention and Water Quality Protection in New York State" (NYSDEC 1996), or (b) conform to a site-specific plan developed by the Whole Farm Planning Program. Most of the farmers select to comply with the requirements of the Whole Farm Plans since they are developed to be site-specific and provide an easier alternative than following the full set of regulations. If a farmer is considered in compliance with his Whole Farm Plan, then they are automatically considered in compliance with the State's Watershed Rules and Regulations. The Whole Farm Plans are a confidential arrangement between the farmer and the Soil and Water Conservation District.

Skaneateles Lake Watershed Agricultural Program

The City of Syracuse contracts with and funds the Onondaga County Soil and Water Conservation District (www.ocswcd.org) to conduct and manage the Skaneateles Lake Watershed Agricultural Program (SLWAP; www.ocswcd.org/index_files/Page771.htm). This program is key to the City's efforts to prevent agricultural operations from polluting the lake, and is designed to be a partnership between the City and the farmers in the watershed. The Soil and Water Conservation District provides guidance to develop and implement plans that protect the lake, as well as funding for some BMPs. Outside funding for additional BMPs is obtained from various grants. While BMPs are paid for by the program, the receiving farmer is required to properly operate and maintain the BMPs. The SLWAP Watershed Agricultural Review Committee provides guidance, advice and direction to the SLWAP staff. As of December 2005, 44 of the 51 farms in the watershed that are eligible to participate in the program are enrolled. In addition to the City, the farmers, and the Soil and Water Conservation District, other partners include the USDA-NRCS, Cornell Cooperative Extension, and the Soil and Water Conservation Districts for the other two counties in the watershed (Cortland and Cayuga Counties).

The Soil and Water Conservation District is a non-regulatory government office, and its personnel are well known to the farming community for helping farmers with a variety of issues. The District's experts work to lend cooperative guidance to the farmers, and provide good continuous support to the City's program. The District receives consistent and sufficient funding from the City to conduct the projects, and the City in turn fulfills part of the requirements for their filtration avoidance waiver.

The Skaneateles Lake Watershed Agricultural Program includes development of Whole Farm Plans, educational programs, and help with design and installation of BMPs for controlling pathogens, stormwater runoff, sediment, phosphorus, and other potential pollutants. In some cases open earthen barnyards have been replaced with concrete pads for containing animal waste. Some barnyards are covered to prevent stormwater from contacting manure on farms with minimal or no filter area between the barnyard and sensitive water courses. Other BMPs include stormwater conveyance systems for farm facilities and grass waterways (grassy swales) for erosion control on sloped crop fields. BMPs conform to the specifications of the USDA-NRCS. Soil and Water Conservation District employees also conduct calibration of manure spreaders and nutrient testing of manure for the farmers in the program. Soil testing is performed to help farmers optimize and reduce fertilizer usage by determining fertilizer needs and developing corresponding fertilizer schedules before application. Manure is used as the primary fertilizer source, and any additional fertilizer needs are satisfied through the procurement of commercial fertilizers.

The Onondaga County Soil and Water Conservation District is currently working with USDA-NRCS to develop and implement Agricultural Environmental Management plans (Whole Farm Plans) for all farms in the Skaneateles Lake watershed, as well as in other specific watersheds elsewhere in the county. As of December 2005, of the 51 farms in the watershed, 36 had Whole Farm Plans completed, and the plans were implemented on 28 of them.

In preparing the Whole Farm Plans for agricultural operations within the Skaneateles Lake watershed and elsewhere in their jurisdiction, the Soil and Water Conservation District relies on guidance from the Agricultural Environmental Management Program started by New York State in 2000 (www.agmkt.state.ny.us/SoilWater/AEM/AEM.html). This program is designed to help New York's agricultural community in its stewardship of the state's soil and water resources, and is a voluntary program through which the Soil and Water Conservation District helps farmers address water quality concerns on their entire farms by coordinating technical and financial assistance. Technical solutions to water quality concerns typically include BMPs that are designed to meet the specific circumstances on each farm. Monitoring of BMP effectiveness is also performed. Some of that BMP performance monitoring is being conducted by researchers at a local university.

One positive aspect of this program is that the farmers participate in the development of their site-specific Whole Farm Plans, and all farmer cooperation in the program is voluntary. If they are involved in the program, farmers may implement and follow the components of the Whole Farm Plan in lieu of being subject to all of the NYSDOH Skaneateles Lake Watershed Rules and Regulations. However, the SLWAP requires an annual review of the Whole Farm Plans, including Nutrient Management Plans and manure spreading schedules, which are updated annually. If a producer is not in compliance with their Whole Farm Plan, SLWAP and the producer work together to develop an agreed upon compliance plan, along with specific completion dates, which outlines all improvements necessary to the get the producer back into compliance.

Another way that farmer participation is maintained at a high level of involvement is through the Watershed Agricultural Review Committee. Seven watershed farmers (and two City representatives) sit on this committee and assist in the formulation and recommendations of program policy, review Whole Farm Plans for reasonableness and consistency, advise program staff on issues of farmer and farm eligibility, and moderate disputes between landowners and SLWAP and encourage watershed farmers to participate in the program. The program also provides incentives for farmers to develop, implement, and follow their Whole Farm Plans. Farms with Whole Farm Plans are provided signs to post, identifying them as participating in the program and willing to apply good conservation measures. Participating farmers can also win Environmental Steward awards from the Skaneateles Lake Watershed Agricultural Program (Figure 7.3).

Since the watershed is fairly small, it's relatively easy for the City's and the Soil and Water Conservation District's employees to get to personally know the individual farmers. This direct contact helps to build a sense of trust between the different parties. Education of both the farmers and the City water department inspectors has been important for maximizing cooperation between the two parties. The City watershed control coordinator and sanitarian are also members of the program review committee.

Example Dairy AFOs and CAFOs

Two dairy AFOs were visited in the Skaneateles Lake watershed in upstate New York, as well as one larger dairy CAFO located nearby but outside the watershed. Both of the two AFOs operate in accordance with a Whole Farm Plan prepared in cooperation with the Skaneateles Lake Watershed Agricultural Program (SLWAP).

The first dairy AFO had approximately 170 total head (with 130 lactating cows), and 1,300 acres of cropland where corn, oats, soybeans, wheat, and alfalfa hay are grown. The cows are housed in a freestall barn year round, and can go outside onto an adjacent concrete pad for exercise when the barn is cleaned. The concrete pad was installed in place of an earthen barnyard to better facilitate manure collection and control runoff. Manure in the barn and on the concrete pad is collected with a skid-steer loader and placed in a manure spreader.

Downhill from the barn and concrete pad is a trench filled with bark to spread runoff from the pad into a subsequent grass filter strip. The bark also acts as a sorptive buffer for runoff from the pad. Most other facilities in this region use gravel trenches instead of bark trenches. After the grass buffer, flow enters a corn field. Silage leachate from the upright silos also goes into the bark trenches. 'Drip lines' (perforated pipes in a gravel collection area) are used to transfer the clean storm water that runs off the roofs for discharge into nearby soil. Rarely does any of the stormwater traverse the whole perforated pipe and result in an above-ground discharge.

Manure is spread once a day (Figure 7.3) in accordance with the nutrient management plan, a component of the Whole Farm Plan. The farm has a three to six week storage capacity for milk house wastewater, which is also spread on the fields. A few years ago this farm installed a flocculator and leach field for treating the milk house wastewater. However, due to persistent problems with lime clogging the drains and the leach field, the system was discontinued and the milk house wastewater is once again applied to crop land. Since this farm does not produce enough manure fertilizer to meet the needs of the crops grown, they purchase supplementary commercial fertilizers. The nutrient management plan has helped them optimize the spreading of manure and thus reduce the amount of commercial fertilizer they need to purchase.

Skaneateles Lake has served as the primary water supply for the City of Syracuse and several outlying communities since 1894

The SLWAP is voluntary for the farmers, and provides guidance and funding to develop and implement WFPs and BMPs

This covered concrete barnyard at a 170-cow dairy AFO replaced an earthen barnyard

Manure is spread daily here, even in the winter, in accordance with a CNMP

Figure 7.3 Photographs of Skaneateles Lake and dairy AFO operations

The second dairy AFO had approximately 80 milking cows housed in a tie stall barn, which are milked in place three times per day. The farm grows corn and hay on 260-tillable acres. Waste from the operation includes cow manure (feces and urine combined), and milk house wastewater that is generated from cleaning the milk storage tank, milking system components, and milk house washdown. In addition to the barn, instead of an earthen barnyard, there is a 150-foot by 50-foot covered concrete pad where the milking cows spend a portion of their time and where manure collection can be easily facilitated and runoff controlled (Figure 7.3). Manure from the pad is scraped up with a front-end loader. The capital costs for the project were paid for by the City of Syracuse, and were administered through the Skaneateles Lake Watershed Agricultural Program. Heifers and dry cows are kept in a small fenced area on an uncovered concrete pad. This pad drains to bark trenches, and the solid waste on the pad is scraped up with a front loader and loaded into the manure spreader. As with the first dairy AFO visited, this facility also uses drip lines at the sides of the farm buildings to help move clean stormwater away.

Collected manure is spread on nearby fields twice-a-day year round, and the milk house wastewater is spread with the second manure load of each day. Cow manure (urine and feces mixed together) and soiled bedding is conveyed by the barn's gutter cleaner and is dumped directly into the manure spreader. Milk house wastewater gravity flows through a floor drain and into a holding tank; from there it is pumped to the manure spreader. This farm spreads all of the manure it produces onto its own crop land. Because of harsh winters, they try to schedule manure spreading during the winter months on flatter fields that are also closest to the farmstead.

This farmer noted there was a "very good working relationship" between the people in the watershed and the City. By no surprise, it helps a lot that the City pays 100 percent of the costs for the agricultural BMPs that the program recommends be installed. The Whole Farm Plan was noted as being helpful in assisting the farmer to better organize his operations.

The third dairy farm visited was an 850-cow dairy CAFO located on 30 acres in a valley south of the City of Syracuse (Figure 7.4). The farm is relatively new, and was built in 2001. The farm operates under a New York State Department of Environmental Conservation (NYSDEC) CAFO permit, and completed an Environmental Impact Statement before construction was approved.

Most of the milking cows are housed in two 436-foot x 98-foot four-row freestall barns, and are herded to the milking center for milking three times per day. Two workers can milk up to forty cows simultaneously. Cows ready to calve, newly fresh cows, and other cows needing some type of medical attention are housed in a separate "special needs" barn. Heifers, calves and dry cows are located at off-site facilities. The raising and harvesting of forages fed at the farm are managed by the farm owner but handled by independent contracting farmers. The farm has access to 1,400-tillable acres.

The farmstead has no discharges to ground water or surface water. All of the waste from the cows is collected using large barn cleaners (sweepers) called alley scrapers and dumped into a below-grade reception pit where it flows via gravity to a single collection tank and is mechanically mixed. The milk house wastewater is collected in a separate system and pumped directly to the farm's long-term storage. Barn manure is pumped to a plug flow anaerobic digester with a detention time of 21 days (capital cost of the digester was over $650,000). The digester microbially produces biogas using some of the remaining food sources in the manure and thus minimizes odor emissions. The biogas is used as an energy source for a boiler, and in turn the heat is used to maintain digester operating temperature (100 degrees F). Excess biogas is burned in a flare. The digester effluent is processed by a liquid-solid separator. Separated solids are recycled on crop land as organic material or sold to the local community as organic matter and the separated liquid is stored in a three-million gallon above-ground storage tank. Groundwater levels in this area are too high (average four feet below the surface) to allow for underground storage. The treated liquid manure is ultimately recycled on the cropland used to grow forage-based crops for the farm in accordance with a CNMP.

850 dairy cows produce a lot of manure. This CAFO uses an anaerobic digester, solid-liquid separation, and long-term storage.

Alley scrapers continuously clears the manure. It flows to a mixing tank and is then pumped to an anaerobic digester.

This anaerobic digester treats manure, and the methane produced generates enough heat to maintain the digester operating temperature.

Solid-liquid separator for digester effluent. The solids are sold as a natural soil amendment. The wastewater is held in a long-term storage tank and applied to fields in accordance with a NMP

Figure 7.4 Photographs of dairy CAFO operations

BEEF CATTLE OPERATIONS (AMARILLO, TEXAS)

City of Amarillo Water Supply

The City of Amarillo, Texas supplies approximately 173,000 people with an average of 40 to 50 mgd (www.ci.amarillo.tx.us/departments/utilities.htm). They obtain 50 to 60 percent of their water supply from a surface water (Lake Meredith), and 40 to 50 percent from 98-production wells in three well fields scattered through four counties. The groundwater supply is used primarily for summer high-demand periods. Much of the City's source water area includes agricultural use, including a relatively high concentration of beef cattle CAFOs as well as crop land where animal wastes are land-applied. The City treats its water using ozonation for enhanced coagulation prior to clarification, dual-media filtration, and chlorine disinfection.

Surface Water Supply

The surface water supply, Lake Meredith, is part of the Canadian River and is located in the Texas Panhandle approximately 45 miles northeast of Amarillo. Lake Meredith is operated by the Canadian River Municipal Water Authority (www.crmwa.com) and supplies water to 11 public water systems, including Amarillo and Lubbock. Approximately 750,000 people are in the service district of these systems and the water is delivered via 322 miles of transmission pipeline.

The source water protection program for Lake Meredith was developed in 2001 and 2002 with funding from the USEPA as a pilot source water protection project conducted by the New Mexico Environmental Finance Center (http://efc.unm.edu). The source water protection strategy included identifying potential pollution sources within 1,000 feet of the lake shore (Zone 1), and a second zone up to one-quarter mile beyond that (Zone 2). Though there are CAFOs in the extended watershed, CAFOs were not identified as a primary source of concern within the primary areas of influence (there are no AFOs located within Zone 1, and only two AFOs within Zone 2). Following inventory of the potential sources of contamination within these zones, the Texas Commission on Environmental Quality completed a source water assessment report (SWAP report) for Lake Meredith in 2003 (Texas Commission on Environmental Quality 2003).

Gas lines underneath and near the lake, transportation routes across the dam and around the lake, sewage lift stations, oil wells and related operations, a lake marina near the intake (concerns with the gas pumps, boat ramp, dock and storage, sanitary system), and storage drums were considered the primary threats to lake water quality. BMPs were listed in the strategy report for each of these potential sources. Also, the Canadian River Municipal Water Authority began the Lake Meredith Salinity Control project in 2001 to intercept brine water leaking into the Canadian River near Logan, New Mexico.

Lake Meredith's source water protection efforts are led by local stakeholders, including representatives from the water systems served by the lake, the Canadian River Municipal Water Authority, and the Retired Seniors Volunteer Program. Other agencies assisting the group include the National Park Service and Texas Railroad Commission.

Groundwater Supply

Most of the supply wells tap water at depths of 400 to 500 feet in the Ogallala Aquifer. One beef cattle CAFO is located within one mile of two of the supply wells, and no other CAFOs are located within five miles of any of the wells. The SWAP report for the City of Amarillo (Texas Commission on Environmental Quality 2003) includes time-of-travel models for each of the supply wells. Wellhead Protection Plans for each well were developed by the Texas Commission on Environmental Quality (www.tceq.state.tx.us).

The City uses regular water quality monitoring and also on-site inspections to help ensure the quality of the groundwater supply. The City analyzes the water from each of their 98 wells once per year, and twice per year for the two wells that are located within one mile of a beef cattle CAFO. No substantial difference in water quality has been observed for the two wells located within one mile of that CAFO and the other 96 wells.

On-site visual inspections are performed every day for active wells, and at least once a month for wells that are seasonally or temporarily out of service. The inspections have proven particularly valuable, as evidenced by one particular incident. A beef cattle CAFO operator located near two of the City's wells applied to the State for an expansion permit, and in doing so noted that there were no public water supply wells within that area. The expansion was approved by the State and developed by the owner in late 2003. This was done despite the two City of Amarillo water supply wells being mapped as adjacent to the feed lot in the SWAP report prepared for the City by the Texas Commission on Environmental Quality and published in May of 2003 (Texas Commission on Environmental Quality 2003).

However, in December 2003, a few days after the cattle were placed into the new pens, City water department employees were doing their monthly inspections of their temporarily out-of-service wells when they noticed cattle living right up next to the well house doors of two of their wells. Apparently the CAFO operator had cut down the City's barbed-wire fence surrounding the wellhead and built cattle pens right around the well houses (the wellheads were located within one-acre tracts owned by the City). The City immediately notified the CAFO operator and the State of the situation, and the problem was quickly resolved by the CAFO operator who purchased the two wells from the City, and paid for construction of two new wells in different locations. There have been other occasions where the City's wellhead inspections have been helpful in finding that cattle or farmers had knocked down the City's barbed-wire fencing used for security around wellheads.

Beef Cattle Operations

Two large beef cattle CAFOs were visited outside of Amarillo, Texas (Figure 7.5). These CAFOs are not located in the Lake Meredith watershed, but are located in the general vicinity of one of the City of Amarillo's three well fields. The first CAFO visited was approximately 35-years old, and had about 30,000-head of cattle. The second was a much newer facility, about 10-years old, and had about 18,000 head. Both facilities have active CAFO NPDES permits for their operations from the State of Texas. In each case, environmental compliance services are provided by the Texas Cattle Feeders Association (TCFA; www.tcfa.org), a representative of which also arranged and led the tours.

At both sites the cattle were separated in groups of similar age and weight and housed in numerous pens. Feed is given two to three times each day to each pen, with records kept of the amount of feed applied and consumed to optimize the feeding regimen.

The waste management at these sites is by nature relatively simple. Manure accumulates in each pen during the life of the cattle there (approximately four-inches of compacted manure). When the cattle are taken away for slaughter the solid manure is removed and hauled off-site (approximately one-half dry ton per head produced) for application on nearby, farmer-owned agricultural land under conditions of the farmer's nutrient requirements for the crop to be grown. The farmer pays for the manure to be land-applied. At one facility some on-site manure composting was also done.

Liquid (stormwater) runoff drains to collection ponds, which are either a natural clay soil deposit or a constructed clay liner (minimum 18-inch compacted clay with hydraulic conductivity < 1.0 x 10-7 cm/sec). The ponds are sized for a 25-year/24-hour storm event, which in the Amarillo area would be about a 5-inch rain storm. The water in the ponds either evaporates, or is pumped to nearby fields as spray irrigation water. At one of the sites, a solids settling basin is located between the bottom of the cattle pens and the stormwater pond. The settling basin provides an initial solids removal zone for water runoff from the cattle pens, and can be cleaned out relatively easy with a front-end, wheeled loader. Drainage ditches uphill of the facilities intercept outside freshwater runoff to prevent it from entering the sites and adding to the liquid effluent managed by the CAFO, and then channels that flow downhill of the facility.

Cattle are housed in corrals and grouped by size, sex, and weight

These cattle are fed two to three times a day. Records are kept of the consumption per pen to optimize the feeding regimen

Corrals slope toward the stormwater collection pond

Solids settling basin precedes the pond; solids are removed with a front-end loader

Corral runoff stored in this clay-lined stormwater pond either evaporates or is sprayed onto nearby fields

Composting is an effective option for handling solid manure

Figure 7.5 Photographs of beef cattle CAFO operations

The soil at these beef cattle CAFO sites is a clay loam, and the manure in the feed pens helps to decrease the soil hydraulic conductivity (the organics in the waste increase the plasticity of the soil). Groundwater quality is periodically analyzed by the Texas Cattle Feeders Association at both sites, and solid manure is analyzed at least once each year. On-site wells provide drinking water for the cattle.

POULTRY OPERATIONS (TULSA, OKLAHOMA)

City of Tulsa Water Supply

Spavinaw Lake and Lake Eucha in northeastern Oklahoma serve as principal water supplies for the City of Tulsa, 17 other cities, and 11-rural water districts. The City of Tulsa obtains approximately half of their water supply from Lake Spavinaw, and the other half from a lake in a different watershed (which is not as affected by nutrient issues). Lake Eucha discharges shortly upstream from Lake Spavinaw and supplies additional necessary storage capacity for the City. Approximately twenty percent of Oklahoma's citizens rely on Tulsa's drinking water. The City of Tulsa treats Lake Spavinaw water with prechlorination, coagulation, sedimentation, granular activated carbon filtration, and chloramination. Current average production from this source is approximately 60 mgd.

Watershed Description

The 415 square mile Eucha/Spavinaw Watershed (OWRB 2002) straddles Oklahoma and Arkansas in the Ozark Highlands and Central Plains ecoregions of the Ozark Plateau. Approximately 60 percent of the watershed is located in Delaware and Mayes Counties in northeastern Oklahoma, and the remainder is in Benton County in northwestern Arkansas. The watershed drains primarily to Lake Eucha, a reservoir of 2,880-surface acres that lies about four miles upstream from Lake Spavinaw (1,638-surface acres) (OCC 2004).

Spavinaw Creek and Beaty Creek are the two primary tributaries to Lake Eucha (Brush Creek, Dry Creek and Rattlesnake Creek are the other tributaries). Spavinaw Creek drains approximately 56 percent of the Lake Eucha watershed basin, while Beaty Creek drains most of the rest. The upstream areas of Spavinaw Creek and Beaty Creek are located in Arkansas.

Land use in the watershed is primarily forest and pasture. Over half of the watershed land is used for agricultural production, of which poultry is the major commodity produced. The watershed includes a relatively large number of chicken farms, related operations, and pasture land where chicken litter has been or is being spread. There are also hog, beef cattle, and dairy cattle operations.

Excessive phosphorus from non-point and point sources in the watershed has led to increased levels of eutrophication and algal growth in the lakes, with resulting decreases in dissolved oxygen and increases in taste and odor causing compounds. During the past 10 years, excessive algal growth has resulted in a steady decline in drinking water quality for northeast Oklahomans, and the City of Tulsa has documented recent avoidance of using both lakes.

Both Spavinaw Lake and Lake Eucha are included on Oklahoma's Clean Water Act 303(d) list as impaired by phosphorus and low dissolved oxygen (2002 listing; OCC 2004). The Beaty Creek tributary to Lake Eucha is also listed by Oklahoma as impaired by pathogens (fecal bacteria), but is not on the 303(d) list in Arkansas.

It is estimated that Beaty Creek and Spavinaw Creek supply approximately 85 percent of the phosphorus entering Lake Eucha (Wagner and Woodruff 1997, cited by OCC 2004). The phosphorus in Beaty Creek apparently largely originates from nonpoint source pollution resulting from land spreading of poultry litter. The phosphorus in Spavinaw Creek likely originates from a combination of both similar nonpoint sources as well as a single major point source (the City of Decatur, Arkansas wastewater treatment plant) (OCC 2004).

While the chicken litter comes from poultry farms in both states, substantially more comes from Arkansas than from Oklahoma (Wimberly and Goodwin 2000). Much of the poultry litter is applied to pasture fields for grazing cattle. The chicken litter is a very good fertilizer for grass, so now many poultry farmers are raising beef cattle (and some dairy cattle) on fields where they grow grass supported by fertilization using their poultry litter.

Watershed Studies and Protection Plans

Numerous studies and programs have been performed to assess the water quality of the lakes and streams in the watershed, identify pollution sources, and develop and implement recommendations for measures to reduce pollution loadings. Select projects are described briefly below.

A Clean Water Act Section 314 Diagnostic and Feasibility Study for Lake Eucha (Phase I Clean Lakes Project; Wagner and Woodruff 1997) conducted by the Oklahoma Conservation Commission (OCC; http://www.okcc.state.ok.us/) on behalf of the City of Tulsa indicated that the watershed was experiencing unprecedented growth of poultry production, and that the dramatic increases in application of poultry litter to pastures in the watershed were the most likely sources of excess nutrients feeding into Lake Eucha. A later study also confirmed this for Lake Spavinaw (OWRB 2002).

The City of Tulsa and the Indian Nations Council of Governments subsequently developed a Watershed Restoration Action Strategy (WRAS) that brought together federal and multi-state agencies and universities for a regionally coordinated approach to addressing water quality degradation in the Eucha/Spavinaw watershed (City of Tulsa and INCOG 1999). Participants included the City of Tulsa, Indian Nations Council of Governments, USEPA, USDA-NRCS, Oklahoma Water Resources Board, Oklahoma Conservation Commission, Oklahoma Department of Environmental Quality, Oklahoma Cooperative Extension Service, Oklahoma Department of Agriculture, poultry integrators and growers, and other concerned stakeholders and citizens.

Accomplishments of the WRAS include establishment of a Watershed Management Team and technical Work Groups, a project management database, an interactive Geographical Information System (available to the public), five scientific studies, a USEPA-approved Quality Assurance Project Plan (QAPP), a poultry litter export hotline, and a common data reference library. Each of these efforts was funded by the City of Tulsa, and they were completed as part of the implementation of the Watershed Restoration Action Strategy. The poultry litter export hotline can be used by chicken producers and farmers to match up available chicken litter to crop fields that could use the manure as fertilizer, including outside of the watershed (http://biosystems2.okstate.edu/poultry/index.asp). Another similar hotline is currently operated by BMPs Inc., a non-profit group supported by the poultry industry. Despite these accomplishments, there was resistance to implementation of some other measures recommended in the WRAS that would reduce phosphorus loadings to the watershed by some of the parties contributing to those loads. The 1999 Watershed Restoration Action Strategy was later effectively superseded by conditions imposed by settlement of the lawsuit described below.

The Oklahoma Water Resources Board (OWRB; http://www.owrb.state.ok.us) completed a 'Clean Lakes Study' ("Water Quality Evaluation of the Eucha/Spavinaw Lake System") in

2002 (OWRB 2002). This scientific report updated the 1997 Diagnostic and Feasibility Study (Wagner and Woodruff 1997), confirmed the previous report's conclusions by identifying phosphorus as the controlling nutrient for the problems in the reservoirs, and described the lake ecosystem conditions in much greater detail and quantified how much phosphorus the lakes could accept. The 'Clean Lakes Study' estimated that a 70-percent reduction in phosphorus entering Lake Eucha would be required to achieve the minimum recommended 45-percent reduction in phosphorus loading to Spavinaw Lake, independent of any other measures (OWRB 2002).

Another study being conducted by a team of scientists in Oklahoma and Arkansas is developing a decision support tool that will be able to analyze impacts of a wide range of land management scenarios on water quality in the watershed (CSREES 2005). The methods and tools for implementing this Nutrient Management Decision and Education Support System should also be applicable to other watersheds.

The Oklahoma Conservation Commission developed a Lake Eucha watershed plan (OCC 2004), which was required by the USEPA before providing a Clean Water Act Section 319 grant to the OCC. The plan summarizes previous and ongoing scientific studies about the watershed and water quality problems, water quality goals and recommended actions, monitoring programs and results, educational programs, implementation schedules for various projects, and a listing of funding needs.

The Oklahoma Conservation Commission has developed several projects that have installed BMPs in the Beaty Creek and Spavinaw Creek subwatersheds, and continue to initiate additional new projects as well as monitoring the effectiveness of the projects completed (OCC 2004). Many of these projects were funded by federal Clean Water Act Section 319 grant funds. The OCC is the state's technical lead agency for implementing Section 319 Nonpoint Source Management Programs. The OCC Water Quality Division is responsible for the prioritization and management of nonpoint source pollution of the state's waters.

Tulsa Lawsuit

Given the resistance of implementation of certain key recommendations of the previously-developed Watershed Restoration Action Strategy by entities contributing to the phosphorus load in the watershed (the poultry industry and the City of Decatur's wastewater treatment plant), the City of Tulsa concluded that the cooperative approach initially taken was failing, and as a result felt that legal action was necessary. Supported with the 'sound science' of the April 2001 draft version of the 'Clean Lakes Study' (OWRB 2001), the City of Tulsa filed a lawsuit against six poultry companies and the City of Decatur, Arkansas in December 2001.

The lawsuit was settled out of court in 2003, and conditions of the settlement were approved by the presiding judge on February 16, 2004. Of particular note is that the 'sound science' supporting the OCC's draft 'Clean Lakes Study' (OWRB 2001) was a key component in the lawsuit being settled in favor of the City of Tulsa. The 1999 Watershed Restoration Action Strategy was effectively superseded by conditions imposed by the lawsuit settlement.

The lawsuit settlement ended the spread of chicken litter to fertilize fields in the watershed until a joint (University of Arkansas and Oklahoma State University) phosphorus risk index was developed to estimate the loads to the watershed and the required reductions to restore reservoir water quality. No chicken litter was allowed to be spread on fields in the watershed throughout 2003, and poultry growers in the basin had to find alternative means of disposal (e.g., transport to and land-spreading outside of the watershed). The poultry integrators also agreed to pay $7 million in attorney fees for the City of Tulsa.

The Eucha/Spavinaw Phosphorus Index developed by the University of Arkansas was a relatively qualitative assessment, while Oklahoma State University's Pasture Phosphorus Management index was more quantitative. Efforts were made to bring the conclusions from these two separate studies into agreement (to develop a 'joint' index), but the parties could not agree and effectively gave that up in fall 2003. Accordingly, the judge issued an Interim Phosphorus Index Order in February 2004 that allowed land spreading of chicken litter that year. The Order required that the application of chicken litter in the watershed be reduced by two-thirds. Previously the loading was approximately 85,000 tons per year, and the required reduction reduced the loadings to approximately 28,000 to 29,000 tons per year in 2004 and 2005. Accordingly, some of the chicken litter produced in the watershed is now transported outside of the watershed for land application. Nutrient Management Plans must also be developed for all those who land apply poultry litter in the watershed.

Point sources are estimated to contribute 23 percent of the phosphorus loading to the watershed (Storm and White 2003). By far the largest phosphorus point source is the City of Decatur wastewater treatment plant (a single large chicken processing plant contributes approximately 84 to 90 percent of the flow and phosphorus loading to the City's 1-mgd wastewater treatment plant). The wastewater discharge previously contained about 7 mg/L phosphorus, and a required phased reduction limited the discharge to 2 mg/L in 2005 and 1 mg/L in 2006. The treatment plant uses alum in its biological treatment to help reach the effluent limitation goals.

The settlement also required that the judge appoint a 'Special Master', an individual who was hired specifically to oversee enforcement of the settlement and to oversee and assist with Nutrient Management Plans. The Special Master serves as a liaison between the court, the lawsuit parties, and, in effect, the poultry growers of the watershed. The Special Master's team of water quality specialists is responsible for collecting soil samples from throughout the watershed wherever landowners desire to spread poultry litter, and then to develop a nutrient management plan that determines the rate at which litter can be spread and other management practices that must be employed (OCC 2004).

In addition, the Special Master's team works closely with a non-profit group (BMP's, Inc.) created by the poultry industry in 2003 to identify additional avenues of reducing the amount of litter spread in the watershed, including coordinating litter transport outside of the watershed, and exploring and implementing alternative uses and markets for the litter. The higher cost of litter transport to outside of the watershed is in some cases partially subsidized by the states (Smith 2005, and http://biosystems2.okstate.edu/poultry/marketing_incentives.asp)

Additional Current and Future Activities

The long-term goals for the Eucha/Spavinaw Watershed are to restore the lakes and streams in the watershed for fish and wildlife beneficial use and to preserve both lakes as a principal water supply for northeast Oklahoma. As of September 2005 there were at least 19 related projects and programs being conducted in the watershed, as well as an active conservation easement work group.

The watershed plan prepared by the OCC in 2004 lists numerous ongoing and planned projects being implemented by a variety of governmental, academic, and non-profit organizations. These projects are designed for implementation of BMPs, monitoring programs, education and outreach programs, and watershed assessment and computer modeling studies. Funding for these projects has been obtained through a variety of sources, including USEPA

Clean Water Act Section 319 grants, the USDA, the states of Oklahoma and Arkansas, the City of Tulsa, the poultry industry, and others.

One particular Clean Water Act Section 319 grant awarded in 2003 is funding education programs, installation of BMPs (riparian areas, pasture management, poultry litter transfer to outside of the watershed to non-nutrient limited watersheds, and septic system replacement), phosphorus assessments, water quality monitoring, and soil sampling to assist in BMP planning. Another grant is paying for validation (data collection) of the Oklahoma soil phosphorus index in the watershed developed by Oklahoma State University (a 2002 grant to OSU).

Like many municipalities, the City of Tulsa can not spend money on projects outside of their political borders. Their costs related to the phosphorus issue in their watershed focus on use of powdered and granular activated carbon (for control of taste and odor causing compounds resulting from algal growth spurred by excess phosphorus loadings), use of algaecide (e.g., copper sulfate) for control of algae in the lake, water quality monitoring of the lakes and tributaries, and funding scientific studies (some via cost-sharing mechanisms). The City of Tulsa and USGS have an ongoing intensive lake and stream monitoring program in the Eucha/Spavinaw watershed (18-monitoring sites), collecting weekly, monthly, and storm-event water quality data and continuous hydrologic data.

In Oklahoma, the Oklahoma Department of Environmental Quality is responsible for Section 303(d) listings and TMDL development. Oklahoma includes both Spavinaw Lake and Lake Eucha on their 303(d) list as impaired by phosphorus and low dissolved oxygen, as well as Beaty Creek for pathogens (from the state's 1998 listing). The Oklahoma Department of Environmental Quality has plans to complete TMDLs for these waters. However, the TMDL program does not require that an upstream state develop or implement a TMDL for a stream if the stream is not listed on that state's 303(d) list. Thus the upstream state (Arkansas) would not be subject to any subsequent TMDLs developed by Oklahoma. However, the Oklahoma Water Resources Board is planning to propose to the USEPA in summer of 2006 a chlorophyll(a) criteria for 'sensitive water supplies' (which includes smaller drinking water reservoirs), and in the fall of 2006 a site-specific phosphorus standard will be proposed for Lake Eucha and Spavinaw Lake with chlorophyll(a) as the response variable. If the USEPA approves Oklahoma's chlorophyll(a) criteria for sensitive water supplies, and the nutrient standard for Lake Eucha and Spavinaw Lake, point and nonpoint sources (e.g., AFOs) of phosphorus in upstream states within the Eucha/Spavinaw watershed will need to comply with any downstream state water quality standard.

BMP Demonstration Farm

The Delaware County office of the Oklahoma State University Cooperative Extension Service (www.delawarecountyext.com) is separately overseeing a demonstration project that is both showcasing agricultural BMPs on a private farmer's land and supporting experiments to study phosphorus runoff from chicken litter applied to grazing fields. The Spavinaw Creek Watershed Demonstration Farm (Figure 7.6) is funded by the OCC (which subcontracted work to the Cooperative Extension Service) using money from a Clean Water Act Section 319 grant.

High phosphorus in Lake Spavinaw caused eutrophication and led to taste and odor problems for the City of Tulsa

Funded by a CWA Section 319, this demonstration farm project showcases agricultural BMPs and runoff field research

Covered concrete cattle feeding shelter located away from creeks. Wells supply drinking water for the cattle

Maintaining natural riparian zones around this creek flowing through the middle of the farm helps to protect water quality

Phosphorus runoff experiments are done using bermed 'mini-watersheds' with stormwater collection and soil sampling

Part of the outreach efforts included this tour of the BMP demonstration farm

Figure 7.6 Photographs of the Spavinaw Creek watershed demonstration farm

The demonstration farm project is trying to change farmers' paradigms of doing things the 'same old ways', and to educate them about more beneficial approaches. This particular site was selected partially because of somewhat difficult conditions such as hilly terrain, poor quality and rocky soil, a creek and riparian zone running through the middle of it, and because it is easily accessible to the visitors it is intended to draw. Once current improvements are completed (e.g., completion of water supply wells for supplying drinking water for the cattle), the farm will be used for raising stock beef cattle, which will arrive after being weaned (approximately 500 lbs) and will be shipped to larger feed lots once they grow to approximately 750 lbs. The BMPs to be included on display include:

- Livestock feeding facility
- Riparian protection zones
- Rotational grazing
- Alternative water sources
- Nutrient management for forage production
- Woodland management zones
- Wildlife management zones

One component of the demonstration farm is a new concrete livestock feed shelter. When not out in the pastures, the cattle can be fed in this covered shelter with a concrete floor that is located a distance from the nearest creeks. The cattle will then frequently defecate on the concrete floor, and the waste can be shoveled and later spread in an environmentally sound manner. With this method a farmer can save money on feed costs by laying the hay on the concrete as opposed to on grass where some of it gets trampled in by the cattle and is thus not eaten. The shelter is approximately 30 x 50 feet, and cost about $15,000. The shelter is also used for receiving and shipping cattle, and for occasional supplemental feedings.

Another suggestion at the Demonstration Farm is to change the crop on the fields to clover in areas where soil phosphorus and potassium levels are already high, as the clover has a lower nitrogen requirement than grasses and would decrease the required maintenance and best utilize the hilly ground. In this case only lime would be required as a soil supplement.

Another suggestion provided to the farmers is to not raise beef cattle in July and August, as it's very hot during that period in this region and the cattle can destroy the fields. The pastures in this area are normally quite good in the fall, and very good in the spring when they can support a much greater number of cattle than at other times of the year.

The Oklahoma State University Cooperative Extension Service conducted a tour of the Demonstration Farm for local farmers and other interested parties in fall of 2005. Of the 65 attendees, approximately 45 of them were local farmers. The farm also hosted a 4-H youth group water quality camp in summer 2005.

The watershed runoff experiments are being conducted by Oklahoma State University and University of Arkansas scientists and include sections of the land which are separated into 'mini-watersheds' by man-made soil berms. Chicken litter of known composition is applied under set conditions, and adjacent control 'mini-watersheds' are also used where no litter is applied. Mechanical runoff collectors are used at the bottom of the 'mini-watersheds', and the runoff is analyzed for phosphorus and other constituents, thus allowing for calculations to be performed regarding the percent that eventually runs off under different conditions. Another section has 21 separate smaller 'mini-watersheds' where different amounts of poultry litter, litter cake (the top section of the litter where most of the manure is), and commercial fertilizer are applied. Soil test results will then be compared between these different plots and the adjacent control plots, and the information will be used to validate a newly-developed Phosphorus Index that takes into consideration factors such as slope, soil type, and proximity to water supplies.

Example Chicken CAFO

An example chicken CAFO was visited in the Eucha/Spavinaw watershed (Figure 7.7). Located in Maysville, Arkansas near the Oklahoma border, the farm has five enclosed shelters that house a total of approximately 126,000-broiler chickens. Three of the houses are 40 x 400 ft (22,000 to 23,000 chickens each), and two of the houses are slightly larger at 42 x 500 ft (30,000 chickens each). The operation is run by a single farmer. Baby chickens are supplied by the corporate integrator, who owns them throughout their life. The integrator dictates many of the conditions for raising the chickens. Once the chickens are grown, the farmer is paid in the form of cents per pound, with the amount being determined by his performance (i.e. feed conversion, mortality) as compared to that of other farmers who provided grown chickens to the integrator in the same week. In other situations, payment may be as simple as paying by the square feet of housing.

The operation is fairly simple as it is totally enclosed and dry, and thus no management of liquid waste is required. Chickens are raised in the houses for approximately six weeks, and eat and drink at will via automatic feeding apparatus and nipple waterers (which help keep the litter drier). Each day the farmer patrols the houses and removes dead chickens, and ensures that the automatic feeding and watering systems are working. Mortality is highest when the chicks are young, and in their last week or two where heart attacks are spurred by the relatively high growth rate of the chickens.

After six weeks, all of the chickens from the five houses are shipped to the integrator's processing facility. A 12- to 14-day period between flocks is used to clean out the houses and as a waiting period to reduce the potential for spreading disease from one flock to the next.

Bedding used in this geographic area is commonly rice hulls, which are relatively expensive. As such, the complete bed of litter (rice hull bedding with feces and urine) is removed only once every year or two (some farmers in the area do so as infrequently as once every six years). Instead of removing all of the litter when each flock of chickens are shipped away, only the litter cake (the upper most layer of the bedding which is largely animal waste) is removed to a covered storage facility with a concrete floor. The litter cake is stockpiled and then spread once a year on the farmer's fields under conditions specified by an approved Comprehensive Nutrient Management Plan. In addition, a couple of neighboring farmers use the rest of the litter. In other geographic areas, farmers can obtain suitable poultry bedding for free, and as such may remove all of the litter after each flock of chickens. This particular farm also contributed litter for experimental use at the Spavinaw Creek Watershed Demonstration Farm described above.

Dead chickens collected from the houses are composted in a series of wooden bins located in a covered storage facility that has a concrete floor. Litter is used as compost material. Multiple compost bins are used for turning the compost and rotating through for ample composting time. Once completely composted, the animal remains are spread on crop land. Composting is the predominant method of disposal of dead chickens in this area. This particular farmer also has a backup incinerator available for disposal of dead chickens. At some other farms, the dead chickens are temporarily stored in a metal container (a small dumpster) or frozen, and then removed by the integrator to be rendered.

This Arkansas chicken CAFO houses approximately 126,000 broiler chickens in five enclosed shelters

This litter cake is removed when the house is empty between each 6-week batch of chickens. Dead chickens are removed daily

Litter is stored in this concrete shelter and then spread once a year on the farmer's fields under conditions of an approved CNMP

Dead chickens are composted in these bins under the side of the litter storage shelter. The composted birds are spread on fields

Figure 7.7 Photographs of chicken CAFO operations

SWINE OPERATIONS (WILMINGTON, NORTH CAROLINA)

City of Wilmington Water Supply

The City of Wilmington, North Carolina obtains their source water from the Cape Fear River, and treats an average of 17 mgd with pre-ozonation, coagulation, clarification, ozonation, dual-media filtration (with granular activated carbon), and chlorine disinfection. The ozonation process was added to reduce DBP precursors and other natural organic matter to help comply with the Disinfectant and Disinfection Byproduct and Surface Water Treatment Rules.

While there are swine, beef cattle, dairy cattle, chicken, and turkey CAFOs in their watershed, the swine CAFOs are the most significant in terms of number, size, impact, and economic importance. The water utility does not coordinate efforts directly with the CAFOs (there are close to 1,000-swine CAFOs within the Cape Fear River basin), and instead relies on the North Carolina Department of Environment and Natural Resources (NCDENR) to monitor the waste management practices of the CAFOs. The utility does monitor source water quality at their intake to assess changes in conditions. This monitoring includes both on-line as well as grab samples for a variety of parameters. They consider their measure of success to be effective, efficient and economical treatment of the source water. Their future goals for source water protection related to CAFO wastes include increased monitoring and source water protection efforts; increased dialogue, communication and partnership with all parties involved; and proactive approaches for remediation of potential problems.

Growth of Swine Operations in North Carolina

Agriculture and its related businesses are North Carolina's number one industry. Among the most significant agricultural producers are the swine and poultry industries. These industries have changed significantly in recent years, generally moving from small farms toward large integrated confinement operations. There was a substantial increase in the number of hog CAFOs in eastern North Carolina in the 1990s. For example, from 1984 to 1994 the number of swine farms in the state dropped from 21,000 to 7,000, while the total number of swine increased. Between 1993 and 1996 alone the total number of swine in the state increased from approximately 4.5 million to 8.5 million (Agricultural Animal Waste Task Force 1996).

This explosion in the growth of swine AFOs and their accompanying on-site waste lagoons resulted in a temporary moratorium (through 1999) on the construction of new swine waste lagoons being issued by the State in August 1997 (General Assembly of North Carolina 1997; NC House Bill 515), which has been extended several times and now goes through September 1, 2007 (per General Assembly of North Carolina, 2003, Senate Bill 593). The purpose of the moratorium was to allow time for counties to adopt zoning ordinances, for scientific studies previously authorized by the state to be completed, and for the state to act on the findings and recommendations of those studies.

That same law also directed the state Department of Agriculture to develop (by May 1, 1998) a plan to phase out the use of anaerobic lagoons and sprayfields as primary methods of disposing of animal waste at swine farms. Nonetheless, lagoon systems and spray fields are presently the dominant treatment method for swine wastes in the state, presumably at least in part due to the substantial political clout and resistance from the state's pork industry to move toward more expensive and advanced waste treatment methods.

Motivation for the moratorium came from considerations of the potential of lagoons to leak and pollute groundwater or surface water, the potential for pollution from spraying of the liquid wastes on pasture and crop fields, and the risk of lagoon berm failures as exemplified by the recent breaking of several lagoon walls during periods of heavy rains, which resulted in large volumes of liquid wastes being discharged directly into nearby surface waters. The largest of these accidents was a major break of the retaining wall of an 8-acre lagoon at Oceanview Farms in Onslow County in June 1995, which spilled approximately 22-million gallons of liquid manure waste into the New River (Agricultural Animal Waste Task Force 1996).

State Regulation and Inspection of Swine CAFOs

A site visit was conducted to the office of the Wilmington Region of the NCDENR (Water Quality Division), which covers seven counties with a total of approximately 680-hog CAFOs (approximately 500 of them are in Duplin County alone). The Wilmington Region does not cover the whole watershed of the Cape Fear River, and there are substantial numbers of hog CAFOs in other nearby counties that are within the watershed but under the jurisdiction of different Regions of the NCDENR (e.g., Sampson County, which is adjacent to Duplin County).

The state of North Carolina regulates waste from animal operations under 15 North Carolina Administrative Code 2H .0217, Waste Not Discharged to Surface Waters (referred to as the ".0200 rules"). The NCDENR is responsible for the permitting process for CAFOs in the state. Any hog farm with at least 250 hogs is required to have at least a State General permit, and any hog farm with 2,500 or more hogs that weigh 55 pounds or greater is required to have a NPDES permit. In the next year or two, there is a possibility that the majority of the North Carolina hog farms that currently are subject to an NPDES permit will be allowed to revert to a State General permit due to the February 25, 2005 ruling by the U.S. 2nd Circuit Court of Appeals (Waterkeeper Alliance Inc. et al. v. USEPA, 399 F.3d 486, 2nd Cir. 2005). For updates on the status of AFO regulations in North Carolina, see http://h2o.enr.state.nc.us/aps/afou/news.htm.

These hog farms are inspected twice per year, once by the Water Quality Division of NCDENR and once by the Soil and Water Division of NCDENR. The latter group tends to provide more technical guidance to the CAFOs, while the former focuses more on determining CAFO compliance with the applicable permits and regulations. The NCDENR Division of Water Quality also conducts visual fly-over inspections after storm events to identify lagoons that have high water levels or may have overflowed, for runoff from the properties, and for active spraying of lagoon effluent after rain or in areas with standing water. The first NPDES permit for a hog farm in North Carolina was issued in March 2001.

Different factors restrict NCDENR's ability to more rigorously monitor CAFOs in the state. As is the case with many state agencies, staffing available for inspections and permit oversight is limited (e.g., the Wilmington Region of the NCDENR Division of Water Quality has three inspectors for approximately 680 swine CAFOs). Political pressure apparently affects the NCDENR's ability to study the influence of CAFOs on groundwater and surface water resources in the state, including the ability to conduct sampling and monitoring programs. The state conducts limited surface water monitoring (and mostly only on larger tributaries), and no groundwater monitoring designed to assess potential impacts from CAFOs.

The NCDENR's on-site hog CAFO inspections focus on potential discharges, the waste collection and treatment systems (e.g., rainfall records and lagoon liquid levels), land application of wastes (e.g., records for sprayer calibrations and spray volumes, analytical results of annual soil sampling), required records and documentation, and other issues such as disposal of dead animals and odor and air quality concerns. Some of the top regulatory requirements that are considered to be directly beneficial to the environment are the required soil and liquid waste analyses (so the CAFO operator knows what they are dealing with), and the mandating of adequate freeboard levels for the lagoons. The soil tests primarily look for metals, pH, phosphorus and potassium. Application of liquid lagoon waste on the fields is sometimes restricted due to soil metals content (especially zinc and copper); once the soil metal standards are reached the facility can't apply any more animal waste on that field. Nitrogen is very mobile so it is not tested here.

The NCDENR inspections also have clearly helped. For example, when these inspections were first started several pipes were found leading from waste lagoons to nearby creeks, but it is now rare for such pipes to be found during inspections. Most of the problems identified during more recent inspections have dealt with record keeping, issues related to land application of liquid waste, and occasionally the identification of lagoons that are filled higher than they are supposed to be (i.e., less freeboard than required).

Example Swine CAFOs

Seven-hog CAFOs were visited in eastern North Carolina (Figure 7.8) in conjunction with scheduled inspections by NCDENR Division of Water Quality personnel. These farms are located in two different subwatersheds, the northeast Cape Fear River and the Black River, both of which drain into the Cape Fear River, which the City of Wilmington uses as their source water. All seven of the farms have the same owner, and are operated by a contract staffing company that provides manpower, utilities, management, and waste management services. The integrator owns the hogs, and provides the food, drugs, and transportation of the hogs. These operations, including related crop land, comprise over 600 acres.

Built in the mid-1990s, these seven farms have a total of approximately 52,000 hogs. One of the farms is a nursery operation, and the rest are all finishers (where the hogs are grown to market weight). There are four to six buildings per site. Some of the finisher buildings are completely roofed ('1224' buildings, named because their capacity is 1,224 hogs), and others are in the 'Missouri style' where only half of each pen has a roof (these have a capacity of approximately 1,320 hogs).

The hogs reside in pens with an area of concrete slats where their feces and urine fall through to a receiving pit channel. Waste in the pits is regularly flushed (approximately five times per day) to an on-site lagoon. The flush water used is recycled lagoon supernatant - lagoon water is pumped back to the flush tanks (adjacent to the hog barns) which store that water for use in flushing the pits.

The lagoons are constructed on 18 inches of compacted clay soil, and provide both settling and anaerobic treatment, as well as storage capacity. The lagoons are built to include some permanent waste and sludge storage, plus six months of waste storage capacity. In addition to that, the level of the liquid in these lagoons (compared to the lowest point of the retaining berm) is regulated to be below a level that includes 12 inches of freeboard (required for structural stability), an additional unused 7.5 inches of storage for a 25-year, 24-hour storm event, and, for lagoons constructed after 1996, another 7.5 inches as a safety factor against heavy rainfall, for a total of 27 inches of freeboard at all times (the depth required for the 25-year/24-hour storm event will vary depending on the location of the farm, and ranges from 5 to 9 inches depending on the location in North Carolina; the heavy rainfall safety factor must be no less than the amount for a 25-year/24-hour storm event). The NCDENR inspector was rigorous in determining the freeboard level at the lagoons at all seven farms visited.

Up to 1,224 hogs are housed in each of the 5 enclosed shelters at this N. Carolina CAFO

This 'Missouri-style' shelter can house up to 1,320 hogs

Drinking water spillage (and thus part of the volume of wastewater generated) is reduced by the use of nipple waterers

Manure falls through slats in the pen floors to a receiving pit channel. The pits are flushed five times a day using recycled lagoon water

Hog manure discharges to an anaerobic lagoon with an 18-inch compacted clay liner

Poor vegetative cover on a lagoon berm can lead to erosion and affect berm integrity.

Figure 7.8 Photographs of swine CAFO operations (continued)

State inspections of hog farms is an important tool. This NCDENR inspector is vigilant checking berms and lagoon freeboard levels

Antiseptic solution is used to clean out the pens between shifts of hogs. This washwater also goes into the lagoon

Stormwater management for roof runoff includes gravel and stone drainage systems

Stormwater ditches move rainwater away from the animal holding areas and lagoon

Natural riparian buffers are left adjacent to drainage channels to reduce sediment and nutrient flow off-site

This drainage ditch has a sluice gate that can be shut in the event of a major lagoon break to stop flow into nearby waterways

Figure 7.8 Photographs of swine CAFO operations (continued from previous page)

Most of the waste handling is liquid, and that is a 365 day-per-year operation. The sludge solids in the lagoons are removed only once every 10 to 15 years. Sometimes it can be problematic for the operator to find a neighboring farm that has use for lagoon sludge solids as free fertilizer since it is quite high in nitrogen, phosphorus, and metals levels.

Lagoon supernatant is disposed of via land spraying as fertilizer on nearby fields of corn, wheat, soybeans, and Bermuda grass under conditions mandated by a required Nutrient Management Plan. At present these NMPs are based on soil nitrogen requirements, but are in the process of being modified to be based on phosphorus requirements using a Phosphorus Loss Assessment Tool (PLAT). All larger hog farms subject to NPDES permits in the state are required to use the PLATs in updating their NMPs.

Stormwater management for roof runoff includes gravel and stone drainage systems. Stormwater ditches direct rainwater away from the animal holding areas and lagoon. Natural riparian buffers are left in place next to farm drainage channels. Some drainage ditches have sluice gates that can be shut in the event of a major lagoon break to stop the wastewater from entering nearby waterways.

Dead hogs are 'rendered', i.e., taken away for disposal by a third party who grinds them up for use in animal feed, as a base product for cosmetic products, and other uses. Dead hogs are placed in a nearby dumpster, or 'dead box' (also known as an ADU container, or animal disposal unit container), which is checked once each day by the third party that removes them from the farms. Elsewhere in North Carolina dead hogs may also be buried or incinerated, and some farms have used composting.

The operator claims to be a proponent of reasonable and consistent regulations. However, they noted there have been lots of regulatory changes over the years, and that shortly after they become accustomed to a set of regulations, the regulations tend to get changed and their staff needs to do more work.

Effects of Hurricane Floyd

Most CAFOs are classified and permitted as nondischarge facilities under the assumption that all waste is contained onsite (including spraying of liquid wastes on nearby fields). However, heavy rain events can create substantial problems with the concept of no discharge. As noted by the operators of the beef cattle CAFOs visited in Texas, the regulations imposed do not account well for the variability in weather often encountered. In the case of swine lagoons, large storms can potentially fill the lagoons, thus requiring disposal of the liquid waste during conditions that are not optimal (i.e., irrigation onto saturated soil).

Severe storm events can potentially create serious pollution problems. For example, Hurricane Floyd brought a massive rain storm to eastern North Carolina on September 9, 1999, with over 20 inches of rain falling within 24 hours in some areas. This storm was greater than a 100-year flood, and wreaked havoc on much of the area. Impacts on some hog CAFOs were severe. The impact of subsequent flooding on confinement buildings, waste pits, and spray fields, as well as large numbers of drowned animals, was well documented. For example, the NCDENR confirmed 46-hog operations with breached or flooded fecal waste pits as a result of Hurricane Floyd's rain (Wing et al. 2002). During Hurricane Floyd, lagoons used by the operator of the sites visited (as part of this project) did not overflow from the rainfall, but instead some creek waters flowed over the lagoons.

Estimates of the number of animals killed in the flooding are as high as 100,000 hogs, 2.4 million chickens and 500,000 turkeys (www.nandotimes.com/nt/images/floyd/floyd063.html).

For a sampling of graphic photos of Hurricane Floyd's impact on hogs farms, see the following (from www.nandotimes.com/nt/images/floyd/index.html).

- www.nandotimes.com/nt/images/floyd/floyd022.html
- www.nandotimes.com/nt/images/floyd/floyd050.html
- www.nandotimes.com/nt/images/floyd/floyd051.html
- www.nandotimes.com/nt/images/floyd/floyd081.html
- www.nandotimes.com/nt/images/floyd/floyd085.html

After assessment of the damage caused by the flooding of Hurricane Floyd, the State of North Carolina implemented a "buy-out" program supported by federal funding to purchase hog facilities and waste lagoons (but not all the adjacent farm land) that were in the 100-year flood plain and were flooded during Hurricane Floyd.

DAIRY CATTLE OPERATION IMPACTS ON GROUNDWATER (CALIFORNIA)

California is the largest dairy producing state in the US. A major portion of California dairy CAFOs are located in low-relief valleys and basins. Large amounts of liquid manure are generated and stored at these dairies. Liquid manure is frequently applied to forage crop lands via flood or furrow irrigation nearly year-round (Harter et al. 2002). Dairies are potentially a major contributor of nitrate and salts to groundwater.

San Joaquin Valley, California

The dairy areas of Stanislaus and Merced Counties have had a long-standing history of nitrate and salt problems in their groundwater. Soils are predominantly sandy and depth to groundwater is typically shallow, thus this area is more susceptible to groundwater contamination from surface activities than other areas.

A research project is being conducted by the University of California at Davis to study the impacts of dairy operations on groundwater quality, and to develop strategies to protect and improve groundwater quality (Harter, Campbell-Matthews, and Meyer 2005). The specific objectives for the research project are as follows:

- Implement a groundwater monitoring program to characterize shallow groundwater quality underneath a number of cooperating dairies in central California
- Develop specific recommendations for improvements in dairy nutrient management that are likely to have significant impact on groundwater quality
- Develop strategies to improve current groundwater quality
- Educate regulatory agency staff and dairy operators about potential pathways of manure nutrients in the subsurface below dairies and about their potential role in safeguarding ground water quality

A groundwater monitoring project was implemented that strategically placed monitoring wells at five dairies in northeastern San Joaquin Valley (Stanislaus and Merced Counties), which are (a) upgradient and downgradient from fields receiving manure water, (b) near ponds, and (c) in the corral area (Harter, Campbell-Matthews, and Meyer 2005; Harter, Meyer, and Mathews 2002; Harter et al. 2002). The 79 total monitoring wells were sampled approximately once every five to six weeks (Harter, Meyer, and Mathews 2002). Typical groundwater levels in this area range from 5 to 15 feet below ground surface. The participating dairies are considered to be well managed and both their operation and geographic environment are representative of many other dairies in the San Joaquin Valley (Harter, Campbell-Matthews, and Meyer 2005).

Monitoring for the array of potential point and nonpoint sources for each dairy was divided into three major source areas: (1) liquid manure water lagoons (ponds), (2) feedlot or exercise yard, dry manure, and feed storage areas (corrals), and (3) manure irrigated forage fields (fields). The monitoring results indicated that nitrate, reduced nitrogen and electrical conductivity (salinity) were subject to large spatial and temporal variability, though the range of observed nitrate and salinity levels was similar on all five dairies. Average shallow groundwater nitrate-N concentrations within the dairies were 64 mg/L compared to 24 mg/L immediately upgradient of these dairies, and average electrical conductivity levels were 1.9 mS/cm (1.9 mmho/cm) within the dairies and 0.8 mS/cm (0.8 mmho/cm) immediately upgradient. Within the dairies, nitrate-N levels did not significantly vary across the three major source areas (dairy management units). However, electrical conductivity levels were significantly higher in corral and lagoon areas (2.3 mS/cm) than in field areas (1.6 mS/cm), thus indicating leaching from the corrals and lagoons. Since fields with manure applied represent by far the largest land area of the dairy, these areas represent the largest potential mass source of contaminants, and thus proper nutrient management during land application will be a key to protecting groundwater quality (Harter, Meyer, and Mathews 2002; Harter et al. 2002). The project is also developing nutrient management practices that will allow operators to fertilize crops with only liquid manure, without significant reduction in crop yield, while minimizing the nitrogen and salt losses to shallow groundwater (Harter, Campbell-Matthews, and Meyer 2005).

RIVERBANK FILTRATION CASE STUDIES

As described previously, riverbank filtration (RBF) systems are particularly well suited for providing better water quality than withdrawal directly from the surface water. Under appropriate circumstances (several dozen feet of unconsolidated, unfractured porous media) reliable RBF systems can remove microbial pathogens, organic contaminants (including taste and odor causing substances and DBP precursors), hormones, endocrine disruptors, turbidity, and other substances. RBF is particularly well suited for *Cryptosporidium* removal (EE&T 2004, AWWA 2004).

The following case studies outline the benefits several utilities have experienced from using riverbank filtration to provide protection against contaminants in surface water, including contaminants that could come from CAFO sources.

Lincoln, Nebraska

The Lincoln Water System (Nebraska) (www.lincoln.ne.gov/city/pworks/water) produces an average of 40 mgd for approximately 250,000 people. They obtain their source water from two horizontal collector wells and approximately 40 vertical groundwater wells. The two horizontal collection wells are located 70 feet below the surface of the river, and are considered as Ground Water Under the Direct Influence (GWUDI) of surface water. In addition to the natural filtration provided by the soil, the utility removes iron and manganese via aeration, chlorination, and filtration, and also via ozonation and filtration. Chloramination is used as the residual disinfectant.

The primary CAFOs in their source water area are beef cattle farms. The SWAP completed for the source water by the Nebraska Department of Environmental Quality identified CAFOs as one of the potential pollutant sources for the river. Since the river flows through three states, the utility has no control over source water protection efforts other than within their limited jurisdiction. As such, the utility relies on the riverbank filtration (and additional treatment) in combination with continually monitoring water quality for contaminants. Given the 70 feet of bank filtration, they do not expect to see much influence from CAFOs on their source water quality.

In support of the value of riverbank filtration, a study by the USGS (in cooperation with the City of Lincoln) showed *Cryptosporidium* and *Giardia* were found in 48 percent and 44 percent of Platte River water samples, respectively, but neither were detected in any of the collector well samples analyzed (Vogel et al. 2005a). Furthermore, *Cryptosporidium* has not been detected in any of the Lincoln Water System's well water samples. In addition, the riverbank filtration is partially effective at removing pesticides such as atrazine. For example, in the springtime after some rain events, relatively high levels of atrazine have been detected in the river, while approximately 40 to 60 percent lower concentration levels were detected in the collector well samples approximately 5 to 7 days later.

Another study by the USGS (in cooperation with the City of Lincoln and the USEPA) examined the occurrence of pharmaceutical compounds in wastewater sites (a cattle feedlot lagoon, a hog confinement lagoon, and wastewater-treatment plant effluent), surface-water sites (Platte River, Salt Creek, and Loup Power Canal), ground-water sites (one collector well and three vertical wells), and drinking-water sites (raw and finished) (Vogel et al. 2005b). Pharmaceutical compounds were detected often in the wastewater-treatment plant effluent, while low-level concentrations of pharmaceuticals were detected in surface and ground water. In contrast, finished drinking water samples did not contain detectable concentrations of pharmaceuticals (other than low levels of cotinine and caffeine). While antibiotics were detected in some of the wastewater samples and twice in Salt Creek, they were not detected in any of the samples from the Platte River or the well field (Vogel et al. 2005b).

Cedar Rapids, Iowa

The City of Cedar Rapids (Iowa) Water Department (www.cedar-rapids.org/water/) produces an average of 38.2 mgd for approximately 130,000 people. Approximately 70 percent of the water demand is associated with grain processing and other large water consuming industries. Prior to 1963 the City treated water obtained directly from the Cedar River. However, taste and odor issues due to elevated algae levels in the river prompted replacement of the surface water intakes with a series of shallow alluvial wells installed adjacent to the river. This system has been expanded over the years and now the City uses 45 vertical wells and 4 horizontal collector wells adjacent to the river (North et al. 2003). The groundwater is treated by the city using lime softening, filtration, and chloramination.

The wells are completed in alluvium at depths up to 72 feet, located 30 to 900 ft from the river. Flowpath studies by the USGS and other agencies have revealed that about 74 percent of the water in the wells originates in the river, 21 percent from the aquifer, and 5 percent from infiltration of precipitation (North et al. 2003; Schnoebelen, Turco, and North 2003; and Schulmeyer and Schnoebelen 1998). These studies also show that some individual wells derive an even greater percentage of their water from the river, up to 90 percent. Subsequent studies have shown that approximately 95 percent or more of the water in the wells originates in the river with the balance from the underlying aquifer and infiltration from precipitation.

The Cedar River drains a major agricultural watershed that also includes some urban areas in Iowa and Minnesota. Approximately 90 percent of watershed area is agricultural, principally corn and soybean production. While there are AFO/CAFO operations in the watershed, there is not a high density. Of those, swine are the predominant CAFOs in the area; there are also beef cattle, dairy cattle, and chicken CAFOs.

The water department worked with state and federal agencies to complete a source water assessment, identifying potential contamination sources in the Cedar River watershed. AFOs/CAFOs were not specifically identified as significant potential contaminant sources in the SWAP. The water utility expects that contributions from individual AFO/CAFO facilities will be diluted before reaching the well field, and that the soil filtration will help to remove contaminants. However, since the area land use is predominately agricultural, the utility is concerned about the cumulative contributions of all agricultural chemicals, especially nitrogenous compounds. The utility's water quality program has confirmed that some contaminants, including nitrate, herbicides and bacteria, enter the Cedar River watershed upstream from the wells.

The watershed upstream of the wells covers an approximately 6,500 square mile area. The city has no jurisdiction and very little "presence" in that area, and thus their source water protection activities have focused on monitoring and educational efforts. The city does meet with Soil Conservation Service and area land owners to discuss water quality issues. These meetings generally focus on nutrient loadings related to row crop activities. Their future goals include improved communication and cooperation with those upstream of Cedar Rapids and will probably not focus on AFO/CAFO wastes.

Water quality challenges include both point and non-point sources, and the greatest contaminant challenge is the elevated levels and increasing trend of nitrate levels. The river routinely has nitrate levels in spring and summer greater than the current MCL of 10 mg/L as N. Monitoring and management of the timing of withdrawals from the wells, coupled with a 2 to 3 mg/L (as N) removal between the river and the wells, has resulted in the City being able to provide water to their customers without exceeding the nitrate MCL (North et al. 2003). In particular, one finding from a four year study conducted by USGS on the wellfield revealed that nitrate levels were 4 to 6 times lower in wells completed in a wetland area adjacent to the river compared to samples from the groundwater or river upgradient of the wetland area (Schnoebelen, Turco, and North 2003).

Other water quality challenges for the river include excessive suspended solids (soil erosion), phosphorus, and fecal coliforms. The 57 mile segment upstream from the City's wellfields has been placed on the State's list of impaired streams due to both nitrate and fecal coliforms. Other problems in the river, especially during the last few years, include atrazine, acetochlor, cyanazine, metolachlor, and other herbicides, along with degradation byproducts of these herbicides. Those herbicides that are currently regulated with primary MCLs (e.g., atrazine) have been detected at some times of the year in the river at levels above the MCL (North et al. 2003; Schnoebelen, Turco, and North 2003).

Riverbank filtration has proven to be highly effective for Cedar Rapids, and RBF water is preferred to treating raw river water as it provides a more consistent and better water quality. The primary benefits of RBF for the Cedar Rapids Water Department are as follows (water quality data summaries are based on average concentrations from analyses performed by the Water Department's in-house certified laboratory):

- Significant reduction in the occurrence of taste and odor events (taste and odor incidents were common when Cedar Rapids drew water directly from the river;

but with RBF they have seen only two short-lived, isolated taste and odor events in the last 15 years)

- Reduction of nitrate levels thus ensuring compliance with regulatory requirements (nitrate concentrations are reduced by approximately 45 percent through RBF)
- Aesthetic enhancements provided by the reduction/removal of algae and compounds that affect taste and odor
- Consistent, low turbidity levels and especially the elimination of turbidity spikes that occur in the river following rain or snow melt events (the well water is much easier to treat than water taken directly from the river). Turbidities are often reduced by 90 percent or more through RBF.
- There is generally 2 to 3 log removal of 2–3 micron-sized particles through RBF, thus suggesting a reduction in microorganisms such as *Cryptosporidium*.
- Some reduction of atrazine concentrations through RBF as evidenced by the avoidance of the extreme spikes that are detected on occasion in the river (however, there are insufficient data to estimate the typical or average percent removal)
- Average TOC reduction of approximately 70 percent (and thus a reduction in DBP precursors)
- Consistent water temperatures and avoiding the extreme high and low temperatures experienced in the river in late summer and winter (again, this makes the well water much easier to treat than water taken directly from the river)

In terms of comparing the costs of RBF versus additional treatment for raw river water, the city does not have specific data to make cost comparisons. However, they believe that the improved water quality benefits from RBF far outweigh the additional pumping, operations and maintenance, and capital costs associated with the alluvial wells. If water was drawn directly from the river, they would expect to incur costs for treatment processes that are not currently necessary with RBF such as presedimentation, carbon filtration and DBP precursor removal.

The utility has also researched the benefits of UV disinfection for adenoviruses, in particular sequential UV-chloramination (Ballester and Malley 2004). Because of the relatively high level of ammonia in the raw water (0.5 mg/L) resulting from agricultural use of the surrounding land, and the expense of breakpoint chlorination and DBP regulations, the water is a good candidate for sequential disinfection (Ballester and Malley 2004), and the city is planning to proceed with the installation of UV disinfection at both of our treatment plants. Capital cost estimates for both of their treatment plants combined are $11 million, and expected operations and maintenance costs are $0.22 to $0.25 per 1,000 gallons.

Mankato, Minnesota

The City of Mankato, Minnesota Water Department (www.ci.mankato.mn.us/utility/treatmain.php3) draws its water from four wells. Combined they supply an average 5.4 mgd to approximately 33,000 people. Approximately 75 percent of the supply comes from a shallow Ranney Collector type well, sixty (60) feet deep, and the remaining 25 percent from three deep wells, cased and open rock hole type, 700 feet deep (two of these are for emergency standby). The Ranney Well has the capacity to pump six million gallons of water per day, and the three deeper wells have capacity to supply an additional 1.1 mgd. The treatment plant uses lime softening, sand filtration, and free chlorine disinfection.

The shallow collector well is located at the confluence of the Minnesota and Blue Earth Rivers, and is considered GWUDI of these rivers. The size of the Blue Earth River watershed contributing water to the City's collector well is approximately 3,500 square miles (Minnesota Department of Health 2003). Agricultural operations dominate the land use within the source water area (the greater Blue Earth River watershed and a portion of the Minnesota River watershed), and the predominant land use is for row crops. The predominant CAFO type in the watershed is swine; there are also beef cattle, dairy cattle, chicken and turkey CAFOs.

The water department worked with the Minnesota Department of Health and other agencies to develop the SWAP for Mankato's water supply (http://156.98.150.16/swa/surfwaterFile/1070009.pdf). According to the source water assessment, the contaminants of greatest concern for the GWUDI supply are pesticides, microorganisms, nitrate, and volatile and synthetic organic compounds. CAFOs were identified as one of the significant potential contaminant sources in the area. Although not required, the City of Mankato intends to use the source water assessment as the basis for development and implementation of a source water protection plan (Minnesota Department of Health 2003). Future efforts to develop and implement a source water protection plan will require the cooperation of various federal, state, and local units of government, and interested parties within the Blue Earth River and Minnesota River Watersheds.

As a means of improving their water treatment capabilities, Mankato is planning to replace their existing sand filters with ultramembrane filters. They claim it has the ability to remove 99.99 percent of *Cryptosporidium*, 99.9 percent of *Giardia*, and practically all bacteria and viruses (City of Mankato 2005). This treatment system will provide an additional barrier to protect against potential microbial contamination from agricultural practices in the source water area, as well as providing treatment benefits related to other contaminants. Water plant construction is expected to begin in spring 2006 and be completed in 2007.

As part of their future source water protection goal of reducing nitrate in their water supply, the utility is also planning to implement a nitrate monitoring program in their source water area. The utility already monitors nitrate levels three times per day to ensure compliance with the MCL, since the levels fluctuate in the collector well as a result of fluctuations in surface runoff. The utility is also interested in means to protect the source water area from contaminants due to releases from CAFO ponds, over-applied land application, and fields with direct flow to streams and rivers.

STAKEHOLDER COOPERATION (PHILADELPHIA, PENNSYLVANIA)

City of Philadelphia Water Supply

The City of Philadelphia Water Department (PWD; www.phila.gov/water) produces an average of 280 mgd for much of the population in the Philadelphia metropolitan area. The City treats their surface water supply with conventional coagulation, clarification, filtration, and chloramination. Their three surface water intakes are located at the bottom of two major rivers - the Schuylkill and Delaware Rivers. About 43 percent of the City's overall water supply comes from the Schuylkill River, and the remainder from the Delaware River.

Source water protection for these rivers is challenging and complex due to the large size of their watersheds and wide variety of land uses and potential contaminant sources. Furthermore, 98 percent of the land area of these watersheds is outside of the political jurisdiction of PWD, making impact on local land use and development decisions difficult. As such, the key to their successful Source Water Protection Program is a watershed approach fueled by partnerships and collaboration.

The Philadelphia Water Department has received several awards in recognition of their efforts for source water assessment and protection, including the USEPA Region 3 Source Water Protection Award (2002), USEPA Clean Water Partner for 21st Century (2003), and AWWA Exemplary Source Water Protection Award (2003, large-size category). Part of the success of the PWD Source Water Protection program is that its vision and mission are supported by a commitment from the top of the organization for substantial resources to work toward achieving those objectives (i.e., PWD Office of Watersheds staff and budget). The Philadelphia Water Department has enthusiastically adopted for itself the role of the region's "watershed champion," accepting the challenge of implementing a comprehensive, environmental approach to water resource management.

Source Water Assessment and Protection Programs

To assess the potential for contamination of their intake water, PWD developed SWAPs for these water supplies under contract with the Pennsylvania Department of Environmental Protection (PWD 2002a, 2002b), as well as for the other water suppliers on these rivers. In all, PWD completed SWAPs for 48 intakes in the Schuylkill River watershed and 8 in the Delaware River. This process helped PWD employees become intimately familiar with the variety of problems facing water suppliers upstream in their watershed. The SWAPs for PWD's intakes identified a variety of contaminants of concern (e.g., pathogens, metals, organic chemicals and salts) originating from both point sources and non point sources. These result from a combination of urban, suburban, industrial, mining, and agricultural activities.

One significant effort was development of a regional early warning monitoring and communication system for water suppliers on both rivers (the Delaware Valley Early Warning System). PWD obtained funding, approached the other water suppliers on the river as well as many other stakeholders, led organizational meetings, and developed the web-based communication system for the early warning system. In combination with the previously-established PWD emergency response plans, the early warning system provides a means to protect the intakes from sudden, major contamination episodes such as large chemical or oil spills.

PWD continued its proactive approach by developing a source water protection plan for the Schuylkill River. The PWD Source Water Protection plan appropriately targeted the contamination sources determined in the SWAP to be most important for potentially impacting the drinking water supply. PWD's source water protection efforts focus on the Schuylkill River watershed for a variety of reasons, including the nature of the source water and its susceptibility to contamination, the smaller size of the watershed compared to the Delaware River watershed, and other factors.

The Schuylkill River is over 130-miles long, includes over 180 tributaries, and drains an area of 2,000 square miles in 11-different counties in southeastern Pennsylvania. It serves as a drinking water source for approximately 1.5-million people. The watershed is approximately 35 percent forested, 34 percent agricultural, and 28 percent developed. The lower watershed is highly developed. Though much of the upper watershed is agricultural, there are only a few farms that meet the regulatory definition of a CAFO.

Given the many other potential pollutant sources and the relatively low number of AFOs and CAFOs in the watershed, and the regulation of CAFOs by the Pennsylvania Department of Environmental Protection (PADEP), these facilities are of relatively low concern. As such,

PWD is more concerned with the smaller unregulated farms, and most of the agricultural source water protection projects are directed toward these facilities. Nonetheless, the PWD program provides an excellent case study in source water protection and developing watershed alliances.

The PWD source water protection program for the Schuylkill River includes various regulatory, structural, and non-structural controls to achieve effective protection. The efforts are quite diverse and include source identification studies and watershed monitoring, piloting of protection technologies, education programs, watershed partnerships, and a variety of implementation projects. While PWD is funding and conducting several projects within their political jurisdiction (the City of Philadelphia), they can not directly fund or conduct projects outside of their jurisdiction. As such, PWD has engaged in an approach to develop a large network of cooperative stakeholder partnerships, primarily through the Schuylkill Action Network (SAN), to coordinate efforts and provide technical information assistance, as well as monitoring and analytical support, for numerous projects further upstream in the watershed.

PWD's Source Water Protection Program uses the following tools to help implement improvement projects outside of their political jurisdiction:

- Combination of the Safe Drinking Water Act and Clean Water Act regulatory frameworks and initiative in a cohesive and productive manner
- A strong emphasis on the development of partnerships with upstream communities to achieve common goals while leveraging outside monies
- The use of technology and technical capabilities to focus limited resources and protection efforts on critical areas that will result in the most cost effective benefit to water supplies
- Coordination and cooperation with other water suppliers in the region to pool resources and efforts towards common goals and reduce overlapping efforts

Overall, through the various techniques listed above, PWD is able to influence the direction of activities upstream of its intakes without any regulatory authority, control, or jurisdiction. This allows for PWD to receive the benefits of upstream protection efforts without having to bear the burden of allocating extensive amounts of PWD's monies and resources into other communities.

On a general level, the draft PWD Schuylkill River Source Water Protection plan incorporates the following major objectives (PWD 2005):

- Establish the Schuylkill Action Network as a permanent watershed-wide organization charged with identifying water quality problems and potential solutions, prioritizing projects, and securing funding sources
- Create a long-term, sustainable fund to support restoration, protection and education projects in the watershed
- Increase awareness of the Schuylkill River watershed's regional importance as a drinking water source
- Initiate changes in policies and decision-making that balance and integrate the priorities of both the Safe Drinking Water Act and Clean Water Act
- Establish a regional Early Warning System for monitoring water quality and improving emergency communications
- Reduce point source impacts to water quality
- Reduce non-point source impacts to water quality

The Schuylkill Action Network

The Schuylkill Action Network (SAN; www.phillywater.org/san) was established in October 2003 to implement the recommendations of the SWAPs. This partnership organization was founded by USEPA Region 3, the Pennsylvania Department of Environmental Protection, and Philadelphia Water Department, and was funded with support from USEPA Region 3. USEPA considers the SAN to be one of their pilot projects for using extensive stakeholder partnerships for source water protection, and they designated a full-time employee to help lead the SAN efforts.

PWD and the SAN have prioritized and initiated numerous projects to address significant point and non-point sources of contamination. Monitoring programs are developed to assess the efficacy of as many of these specific projects as practical. The SAN consists of four technical workgroups, each charged with addressing one of the four major sources of contamination, including Abandoned Mine Drainage, Agricultural Activities, Urban and Suburban Stormwater, and Pathogens and Regulated Discharges. Support teams were also formed to address ancillary and support issues such as planning; data collection, analysis, and management; education and outreach; and land acquisition (the Watershed Land Protection Collaborative).

The concept of the SAN to serve as a "one-stop shop" for stakeholders to find funding sources is an excellent approach. In addition to the financial commitment made by PWD for projects within their watershed, numerous grants have been received by various watershed organizations that are members of the SAN for complementary protection efforts. This includes a $1.15-million grant received from USEPA in April 2005 that is funding 35 different SAN projects (the grant is being administered by the Partnership for the Delaware Estuary). The SAN is currently working toward establishing a plan for securing long-term funding to support future efforts. PWD and the SAN also support numerous educational and informational programs that serve to increase awareness of the Schuylkill River watershed's regional importance as a drinking water source.

Agricultural BMP Projects

The goal of the SAN's Agriculture Workgroup is to address the 258 miles of streams impaired due to high loadings of sediment, nutrients, and pathogens from agricultural practices. The workgroup includes a large and diverse group of stakeholders, including representatives from five county Conservation Districts, four water suppliers (including PWD), local and state Farm Bureau offices, a local Grange, various NRCS offices, the Pennsylvania Department of Environmental Protection, USDA, USEPA Region 3, and assorted planning commissions, cooperative extension, and conservancy groups. Approaching the relevant landowners and agricultural interests is much more fruitful when done by the agricultural partners within the SAN rather than by the 'big city' government located many miles downstream.

The Agriculture Workgroup's efforts included an investigation to identify and prioritize critical land areas for further work. The Workgroup identified all farm parcels in the watershed that contain an agriculturally impaired stream, and then prioritized those 800 farm parcels using a statistical tool to rank farms based on stream order, acreage, stream length, and other criteria. A field survey was performed of the top-ranked farms determined during prioritization, and tours were taken of some of the key farm sites (including a dairy CAFO) where BMPs were installed as part of Clean Water Act Section 319-funded programs.

The Agriculture Workgroup selected three subwatersheds in which to focus possible implementation of agricultural BMPs. Funds were obtained by different organizations in the

SAN for a variety of projects, including development of conservation plans for farmers on impaired streams, riparian buffer assessments, supplies for volunteer water quality monitoring programs, and implementation of various BMPs including streambank fencing, cattle crossings, riparian buffers, stream bank stabilization, infiltration step pools, invasive species removal, and removal of multi-flora and the planting of native trees. Funding for these projects came from a variety of sources including the NRCS, State of Pennsylvania (Growing Greener Grants), and USEPA (Clean Water Act Section 319 grants and a Targeted Watershed Initiative Grant).

In addition to those projects being completed through the SAN, the PWD has also been involved in several agricultural-related projects within the city limits. These include the Fox Chase Farm, Courtesy Stables, Monastery Stables, and Saul High School projects. While Fox Chase Farm is in the Pennypack Creek Watershed, which drains to the Delaware River, these other projects are all within the Wissahickon Creek Watershed, which drains to the Schuylkill River.

The Fox Chase Farm Riparian Buffer project was implemented in May 2002, and included the establishment of a 1.85-acre riparian buffer that included 400 trees and 735 shrubs for a width of 45 feet on either side of the stream. Streambank fencing was also installed along with a cattle crossing. Where the tributary originates on the farm, fecal coliform and *E. coli* counts were reduced more than 90 percent as a result of these measures.

The Courtesy Stables and Monastery Stables projects use runoff diversion and infiltration measures to hinder the flow of stormwater runoff through manure-laden paddock areas and to divert the flow to a final detention basin to allow settling before discharging to the receiving stream. In the case of the Courtesy Stables project, the final basin includes wetland vegetation to provide treatment prior to discharge. In the case of the Monastery Stables, runoff enters a basin which overflows to a holding tank which allows further settling prior to being discharged. Both projects are now completed but post-implementation monitoring has not yet been conducted.

Saul High School is an agricultural school in the city that includes horses, cows, and sheep. Runoff from the school's pastures created an eroded gully which transports the pasture runoff to the Wissahickon Creek, which feeds into the Schuylkill River. This project, planned for construction in spring 2006, includes a riparian buffer and a series of settling ponds to encourage infiltration and reduce pathogens, nutrients, and sediments entering the Wissahickon from this location.

Benefits of Stakeholder Partnerships

In summary, the Philadelphia Water Department's source water protection efforts are particularly challenging in that their raw water intakes are located at the bottom of two major rivers, and most of the land area of these watersheds is outside of their political jurisdiction, making impact on local land use and development decisions quite difficult. Nonetheless, the water utility has risen well to this challenge through substantial outreach to local governments and many other stakeholders. Furthermore, PWD has turned some of this disadvantage into an asset by outreaching to a greater number of partners to share and leverage resources to address various common water resource issues, and by developing a sustainable and coordinated framework (e.g., the Schuylkill Action Network) to achieve true environmental outcomes and benefits. In this manner, PWD does not need to create an expensive and resource-intensive overarching program that controls all aspects of water resources in their watersheds. Instead it allows them to work within and support existing frameworks and initiatives to influence upstream solutions without extensive resource investments.

CHAPTER 8
STRATEGIC WEB-BASED INFORMATION CENTER

As a complement to this report, an electronic information center was developed and is available on the project website at www.eetinc.com/cafo. The electronic information center can help a utility to evaluate their site-specific situation and identify potential cost-effective proactive measures. The website includes some of the information from this report, and also provides organized and instant access to a vast quantity of information available on the internet.

Water utilities can use the website to identify information sources and agency contacts that can help them identify CAFO locations in their source water area. While it was impractical to make this an exhaustive or "up to date" database, links are provided to state and federal maintained databases. Once CAFOs are identified in a source water area, a utility needs to determine which CAFOs, if any, present a risk to their source water quality. If a utility has concerns about a specific CAFO or group of CAFOs, it should establish a dialogue with the CAFO(s), either directly or through agricultural or regulatory organizations. The nature of the CAFO and its wastes, as well as the BMPs that are in place, should be explored. This information will allow the utility to investigate the effectiveness of these BMPs, and possibly identify reasonable alternatives that may be more beneficial.

It is expected that the most beneficial output from the website for many utility users and researchers will be references to other available information sources for a particular issue, including links to information sources available on the internet from USDA, USEPA, USGS, AWWA, AwwaRF, universities, state agricultural extension services, and other publicly available information. State, federal, and CAFO organization contact information is also provided to allow utilities to gain quick access to information.

CHAPTER 9
APPLICATIONS POTENTIAL

Water utilities are encouraged to identify the location and characteristics of AFOs and CAFOs in their source water areas, determine if these facilities pose a risk to source water quality, develop a prioritized plan of action for addressing key issues, and implement the components of the protection plan. The final report and web-based information center tool are intended to provide background information and guidance to help utilities assess potential impacts of AFOs and CAFOs on their source water quality, and for developing appropriate, successful and affordable courses of action in cases where AFOs are considered to be of concern.

This report provides water utilities with the background necessary to understand the basics of animal feeding operations, the quantity and characteristics of wastes produced, the potential effect of these operations on drinking water source quality, typical waste management strategies for AFOs, and measures utilities can take to help protect the quality of their source water.

Proactive strategies that the animal producers may employ are outlined and compared, including BMPs for animal waste reduction, management, treatment and disposal. Development of cooperative relationships between the utilities and the agricultural community and other appropriate stakeholders is usually an effective means of leading toward assessment and action remedies. Concerned water utilities should work with AFOs and CAFOs, and also with their regulators and other agricultural stakeholders, to make sure BMPs are employed to limit both the generation of contaminants (e.g., reduce hormones, antibiotics, excessive nutrients and other contaminants in feed supplies and supplements) and contaminant releases from the AFO processes (e.g., on-site storage and treatment, optimized land application of waste, etc.).

Numerous strategies that water utilities can take are discussed. Utilities are provided with a collection of alternative strategies to choose from, which are characterized so utilities can select the alternative most in keeping with their needs, capabilities, and goals. Alternatives include various forms of monitoring, source water protection, intake management, pretreatment options, and effective water treatment operations they can employ to provide protection for source and treated water quality.

Measures for protecting against both chronic contamination and larger-scale accidental releases from AFOs are described. Microbial source monitoring techniques can accurately assess the sources of pathogenic contamination in water supplies, thus allowing utilities to appropriately prioritize their actions toward the most important potential contamination sources. Water quality monitoring can be used to verify effectiveness of currently used BMPs, and to identify any accidental releases. Early warning systems can provide timely information about rapid changes in water quality so that intake management and treatment practices can be manipulated as necessary. Utilities can use this guidance for proactive source protection, pre-treatment, and treatment operations. Sources of funding for various source water protection projects are also identified and discussed.

CHAPTER 10
FUTURE RESEARCH NEEDS

Consideration of the information and experiences obtained during this research has led to the following recommendations for future research and programs that would help water utilities better understand and assess the potential impacts of AFO operations on source water quality, as well as identify appropriate and effective courses of action for AFOs and utilities to take:

- Further review available data and scientific assessments of the effects of AFOs on source water quality and how they may impact subsequent treatment for drinking water
- Assess nutrient, pathogen and pesticide fate and transport through soils, to groundwater, and to surface waters from AFOs and related farms
- Improve methods for determining the absolute and relative contribution of different sources of contaminants to source waters, including from AFOs, other agricultural operations, and urban, suburban, industrial, and other land use areas
- Improve methods for modeling of the effects of agricultural BMP implementation on source water quality
- Improve methods for obtaining field parameters needed for watershed and subsurface transport model development, calibration, and verification
- Further investigate the levels of emerging contaminants such as hormones and pharmaceuticals that can be found in various animal wastes, whether releases of those chemicals from AFOs are contaminating drinking water sources, and whether that contamination is at a level of public health concern
- Evaluate design and regulatory criteria to prevent or mitigate uncontrolled discharges from CAFOs during storm events more severe than those of 25-year recurrence interval
- Examine the correlation between *Cryptosporidium* and *Giardia* occurrence in source waters and the presence of AFOs and CAFOs in the watershed
- Study innovative in-lake management control techniques for lake restoration due to excess nutrient loading
- Develop methodology for determination of optimal buffer widths for treatment of various pollutants
- Examine the effect of manure constituents on the taste of drinking water

USEPA compiled a list of research needs associated with CAFOs as part of their Risk Assessment Evaluation for Concentrated Animal Feeding Operations (USEPA 2004b), and included the following priorities to address:

- Contaminant evaluation and quantification
- Development of new methods to rapidly and inexpensively measure contaminant levels
- Development of new methods to identify sources of contaminants in the environment, including isotopic and microbial contaminant source tracking
- Treatment process research, from bench-scale to field-scale, to find cost-effective processes and optimal treatment conditions for control of different contaminants

- Effects of storage and treatment processes on inactivation of pathogens
- Fate, transport and effects of AFO contaminants in the environment
- Effects of management practices
- Identify alternate uses for wastes and salable products that can be generated from waste streams
- Odor and airborne particle production and control

CHAPTER 11
RECOMMENDATIONS FOR WATER UTILITIES

Presented below are some of the more notable recommendations developed during this study for drinking water utilities to assess potential impacts of AFOs and CAFOs on their source water quality, and for planning and implementing appropriate, successful and affordable courses of action for protecting source waters from contamination from CAFO facilities and other farms.

- Water utilities are encouraged to perform comprehensive source water assessments. All potential waste sources should be considered for assessment and prioritization; in addition to CAFOs, the rest of the agricultural sector (including poorly managed farms of any type) and any other land uses in the source water area should be considered.
- All source waters should have had a SWAP report prepared under requirements of the 1996 SDWA. Depending on the quality and thoroughness of that SWAP report or other related assessments, utilities may wish to conduct additional study and monitoring.
- At minimum, water utilities should know if there are CAFOs in their source water area, and whether these operations have an adverse impact on source water quality.
- For utilities concerned with actual or potential pollution from AFOs, they should identify the location and characteristics of AFOs and CAFOs in their source water areas, determine if these facilities pose a risk to source water quality, develop a prioritized plan of action for addressing key issues, and implement the components of the protection plan.
- As with any source water protection program, utilities should use a multi-faceted approach. This can include source water assessment, monitoring, stakeholder alliances, outreach and education, and encouraging effective BMPs at nearby priority AFOs.
- Periodic reevaluation of source water assessments and of source water protection program action priorities and success will help ensure the program is achieving optimal benefits.
- AFO operators will be more willing to cooperate if they can be shown through sound science that their operations and actions have an adverse impact on a water source, and that if they take appropriate corrective measures it will produce reasonable and measurable improvement in source water quality. A sound source water assessment that analyzes the relative contributions of potential pollutant sources in a source water area can go a long way toward providing the scientific evidence necessary to effectuate change in land use practices. Utilities can help themselves by using the right experts and expertise to perform sound scientific assessments.
- Hydraulic and contaminant transport modeling of watersheds or groundwater can be used to more fully estimate potential impacts from priority contaminant sources. Modeling can also be used to assess the anticipated general effect of possible BMPs and prioritize actions. Assessments and data showing agricultural

impacts on source water quality should be locally and regionally based in order to be more convincing to local and regional agricultural interests.

- Monitoring of source water quality is important for identifying potential pollutant problems and sources (to target protection efforts), and for verifying the effectiveness of source water protection programs (including monitoring the effectiveness of BMPs at AFOs). Many monitoring programs target both dry and wet (storm event) sampling of their source water and upstream locations, as the primary pollutant source may vary depending on the amount and recent frequency of rainfall and runoff, the amount of fixed discharges (e.g., public sewage treatment plants), and other factors.
- For some utilities, monitoring combined with treatment may be a sufficient approach. Monitoring programs are used as a primary intake water protection measure by many utilities that have no other means of controlling outside activities in their source water area, including utilities drawing source water from major rivers with extensive watersheds.
- Concerned water utilities should work with AFOs and CAFOs, and also with their regulators and other agricultural stakeholders, to make sure BMPs are employed to limit the generation of contaminants and to limit and control contaminant releases from the AFO processes. Water utilities can support development and implementation of whole farm plans and comprehensive nutrient management plans.
- Establishing good communication and collaboration between the water utility, farmers, regulators, and other key stakeholders is essential for effective source water protection, and can not be overemphasized. One of the most effective means for concerned water utilities to approach agricultural interests is through cooperation with entities that already have relationships established with the local farmers, such as county soil and water conservation districts, cooperative extension offices, US Department of Agriculture (USDA) Natural Resources Conservation Service (NRCS) offices, and other related organizations. Working through local agricultural and watershed stakeholder organizations and coalitions generally builds trust better than when utilities initially approach farms and the agricultural community on their own. Other state and federal agencies may also be willing to provide financial and technical assistance to help alliances improve water quality. Water utilities should take advantage of these resources, rather than try to do all the work on their own. This is both a more effective and more economical approach.
- Another group of potential partners not to overlook are other water systems and municipalities in the source water area that would also benefit and could contribute additional resources and responsibilities for assessments and corrective measures in the watershed. These other utility partners should be involved in discussions with farmers, regulators, and other stakeholders that are part of your team.
- Water utilities should start their approach to the local agricultural community very non-aggressively. Utilities need to demonstrate understanding of the AFO operators' concerns. Trust must be established before changes in farm operational practices are likely to be made. When utilities push too hard and too fast, they sometimes meet with greater resistance. Every watershed is different, and utilities need to learn who the effective players are in their area and what lines of

communications are most beneficial. It can be beneficial to work with key producer leaders to develop a relationship with their industry. Often other producers will follow suit based on what they see the leaders doing.

- Always try to find win-win scenarios with local farmers. Approach farms asking what you can do to help them, as opposed to telling farmers what to do. Emphasize the benefits that both their farms and the environment will realize from the suggested improvements in farm operations. Realize that the first question will usually be "who will fund the suggested BMP improvements?" Providing incentives such as funding for BMP implementation or identification of measures that can increase farm profitability can be very useful.
- Funding is a common obstacle to improvement of agricultural operations for protecting drinking water source quality. Many successful case studies have shown some type of cost-sharing to be effective. Proactive water utilities can work with farmers and other stakeholders to find funding for a variety of agricultural source water protection projects. Knowledge of potential funding sources, as well as good relationships with stakeholder organizations that are experienced at gaining related funding, can greatly help a water utility to gain cooperation of the farms in their source water area.
- Individual education at the farmer's facilities can be highly effective, as it allows for direct assessment and discussion about practices conducted at that farm, and many proactive water utilities have programs for this. Farmer involvement in development of their Whole Farm Plans and Nutrient Management Plans leads to a better understanding and acceptance of what they can and should do. Some water utilities, such as Tulsa (Oklahoma) and Springfield (Missouri), support demonstration farms that are used to showcase various agricultural BMPs, and educate farmers as to more beneficial approaches.
- One water utility strategy that is particularly well suited to situations where AFO wastes and other contaminants are possibly present in surface waters is to replace traditional surface water intakes with wells and other subsurface collection systems located adjacent to the surface water source. These systems, typically referred to as riverbank filtration, use the subsurface media between the source and groundwater well to remove a variety of contaminants, including microbial pathogens, organic contaminants (including taste and odor causing substances and DBP precursors), hormones, endocrine disruptors, turbidity, and other substances.

APPENDIX A
WEBSITES RELATED TO CAFOS AND CAFO WASTES

USEPA:

- USEPA CAFO Rule http://cfpub.epa.gov/npdes/afo/cafofinalrule.cfm
- From the Federal Register. National Pollutant Discharge Elimination System Permit Regulation and Effluent Limitation Guidelines and Standards for Concentrated Animal Feeding Operations (CAFOs); Final Rule (40 CFR Parts 9, 122, 123, and 412). Volume 68, Number 29, Pages 7175-7274, February 12, 2003
- USEPA Rule extending certain CAFO operator compliance deadlines for the CAFO Rule in order to allow USEPA more time to comply with the Court's decision ("Revised Compliance Dates for National Pollutant Discharge Elimination System Permit Regulation and Effluent Limitation Guidelines for Concentrated Animal Feeding Operations", Federal Register, Vol. 71, No. 28, February 6, 2006; available at http://www.epa.gov/fedrgstr/EPAWATER/2006/February/Day-10/w1240.htm)
- Proposed revisions to the 2003 CAFO Rule signed by the USEPA Administrator on June 19, 2006 and issued for public comment to more broadly to address the ruling of the Second Circuit Court of Appeals; available at http://cfpub.epa.gov/npdes/afo/aforule.cfm
- USEPA's National Agriculture Compliance Assistance Center: (http://www.epa.gov/agriculture) A fair amount of information on animal production may be found on this website (see http://www.epa.gov/agriculture/anafopro.html). The website also has links to related information on BMPs, research, regulations, and other pertinent topics
- AFO Virtual Information Center (http://cfpub.epa.gov/npdes/afo/virtualcenter.cfm). The AFO Virtual Information Center is a tool to facilitate quick access to livestock agricultural information in the US. Includes links for programs and information in each state
- USEPA NPDES permit guidance documents (http://cfpub.epa.gov/npdes/afo/info.cfm#guidedocs)
  - USEPA NPDES website for AFOs (http://cfpub.epa.gov/npdes/links.cfm?program_id=7):
  - The Producers' Compliance Guide for CAFOs (http://www.epa.gov/npdes/pubs/cafo_prod_guide_entire_doc.pdf) describes USEPA's regulations for concentrated animal feeding operations (CAFOs). The *Producers' Guide* includes sections that describe USEPA's CAFO regulations, who they affect, what they require, and what assistance is available. EPA 821-R-03-010. November 2003. Additional description available at http://cfpub.epa.gov/npdes/afo/compliance.cfm#copy#copy
  - USEPA 2003. NPDES Permit Writer's Guidance Manual and Example NPDES Permit for Concentrated Animal Feeding Operations. December 31, 2003. EPA-833-B-04-001 http://cfpub.epa.gov/npdes/afo/info.cfm#guidedocs

- Example NPDES General Permit for Concentrated Animal Feeding Operations. http://www.epa.gov/npdes/pubs/cafo_example_permit.pdf
- NPDES Form 2B Application (http://www.epa.gov/npdes/pubs/cafo_fedrgstr_form2b.pdf). Application for permit to discharge wastewater from concentrated animal feeding operations (CAFOs) and aquatic animal production facilities. EPA Form 3510-2B (12/02). OMB No. 2040-0250
- Managing Manure Nutrients at Concentrated Animal Feeding Operations (http://www.epa.gov/npdes/pubs/cafo_manure_guidance.pdf) (USEPA 2004a; EPA-821-B-04-009; 12/01/2004). Technical guidance designed to supplement the NPDES Permit Writers' Guidance Manual and Example NPDES Permit for Concentrated Animal Feeding Operations. This guidance provides additional technical information for owners, operators, technical service providers, consultants, and permit authorities on how to carry out USEPA's revised regulatory requirements for NPDES permitting of CAFOs. It also provides information on voluntary technologies and management practices that may both improve the production efficiency of CAFOs and further protect the quality of the nation's waters. This document assumes that readers have a basic understanding of the CAFO regulations. Also available at http://cfpub.epa.gov/npdes/docs.cfm?program_id=7&view=allprog&sort=name

• CNMP Watch (http://www.cnmpwatch.com/) is a good web source for Comprehensive Nutrient Management Planning (CNMP) information. CNMP Watch is a project of the National Association of State Departments of Agriculture Research Foundation (NASDA). The project is funded by the USEPA. Contains links to relevant information by state
• USEPA Program and Regional Offices' Links (http://cfpub.epa.gov/npdes/linkresult.cfm?program_id=7&link_category=7&view=link) - Provides links to other USEPA Program Office web sites that are related to the NPDES permit program
• Other USEPA Specific Programs and Initiatives (http://cfpub.epa.gov/npdes/linkresult.cfm?program_id=7&link_category=1&view=link) - Provides links to specific USEPA programs and initiatives related to the NPDES permit program
• Livestock Environmental Stewardship Brochures. Pictorial description of environmental stewardship for animal feeding operations:

  - Beef www.epa.gov/agriculture/beef.pdf
  - Dairy www.epa.gov/agriculture/dairy.pdf
  - Poultry www.epa.gov/agriculture/poultry.pdf
  - Swine www.epa.gov/agriculture/swine.pdf

• Managing Livestock, Poultry, and Horse Waste to Prevent Contamination of Drinking Water. Source Water Protection Practices Bulletin. USEPA. 2001. EPA-916-F-01-026. This fact sheet addresses some source water contamination prevention measures related to livestock, poultry, and horses. http://www.epa.gov/safewater/protect/pdfs/livestock.pdf.
• Managing Agricultural Fertilizer Application to Prevent Contamination of Drinking Water. Source Water Protection Practices Bulletin. 2001. EPA-916-F-

01-028. This fact sheet addresses some source water contamination prevention measures related to agricultural fertilizer application. http://www.epa.gov/safewater/protect/pdfs/fertilizer.pdf

- Managing Manure Nutrients at Concentrated Animal Feeding Operations (USEPA. 2004a) - technical guidance implementing CAFO Rule requirements, including land application methodology. http://www.epa.gov/npdes/pubs/cafo_manure_guidance.pdf or http://cfpub.epa.gov/npdes/docs.cfm?program_id=7&view=allprog&sort=name
- Risk Assessment Evaluation for Concentrated Animal Feeding Operations (USEPA 2004b). EPA/600/R-04/042. Discusses the potential environmental problems associated with CAFOs, focusing on beef, dairy, swine and poultry operations. http://www.epa.gov/ORD/NRMRL/pubs/600r04042/600r04042.pdf.
- USDA/USEPA Unified National Strategy for Animal Feeding Operations. March 9, 1999. Sets forth a framework of actions for USDA and USEPA under existing legal and regulatory authority to reduce impacts to water quality and public health from improperly managed animal wastes. Provides information to NPDES permitting authorities, owners and operators of animal feeding operations, and the general public on how to implement the Clean Water Act CAFO regulations, including: 1) information to help permitting authorities ensure that NPDES permits conform to the CAFO regulations and 2) general information on Clean Water Act and NPDES requirements that USEPA will consider when reviewing the adequacy of state NPDES permits for CAFOs. http://cfpub.epa.gov/npdes/afo/ustrategy.cfm
- Alternative Technologies/Uses for Manure (Draft Report). This document is a compilation of alternative uses for manure from AFOs, including conversion of manure to value-added products (e.g., as fertilizers, soil amendments or feed additives) and as an energy source. http://www.epa.gov/npdes/pubs/cafo_report.pdf
- State Compendium: Programs and Regulatory Activities Related to Animal Feeding Operations: includes a compilation of descriptions of AFO-related state regulatory programs and initiatives, with a specific focus on the use of permits or similar mechanisms. Profiles are included for each state addressing background information, lead regulatory agency, state regulations regarding AFO/CAFOs, types of permits, permit coverage, permit conditions, enforcement information, state voluntary programs, and additional state-specific information. Prepared in 2002 by Tetra Tech, Inc. (Fairfax, VA) for the USEPA Office of Wastewater Management (Washington, D.C.) http://www.epa.gov/npdes/pubs/statecom.pdf.
- The ASCE/USEPA International Stormwater BMP Database includes numerous reports of stormwater BMP efficiency measurements (http://www.bmpdatabase.org.)
- CAFO Final Rule Supporting Documents: A single website that contains links to the various supporting documentation used for development of the CAFO Rule (http://cfpub.epa.gov/npdes/afo/cafodocs.cfm). Includes links to the following reports:
  - Technical Support Document for the Revisions to the National Pollutant Discharge Elimination System Regulations for Concentrated Animal Feeding Operations. Supporting document for the CAFO Final Rule. Support materials for the final NPDES CAFO regulations including

affected entities and state costs (EPA-821-R-03-002). http://www.epa.gov/npdes/pubs/cafo_tech_support.pdf.

- Economic Analysis of the Final Revisions to the NPDES and Effluent Guidelines for CAFOs Supporting document for the CAFO Final Rule. Economic Analysis (EA) summarizes USEPA's analysis of the estimated annual compliance costs and the economic impacts that may be incurred by affected operations that are subject to the final revisions (EPA-821-R-03-002). http://cfpub.epa.gov/npdes/afo/cafodocs.cfm.
- Development Document for the Proposed Revisions to the NPDES Regulation and the Effluent Guidelines for CAFOs. Supporting document for the CAFO Final Rule. This Technical Development Document (TDD) presents the legal authority, data, and summaries of the analysis that the USEPA used to revise regulations for the CAFO industry (EPA-821-R-03-001). http://cfpub.epa.gov/npdes/afo/cafodocs.cfm

    - Chapters:

        - http://epa.gov/ost/guide/cafo/pdf/DDChapters1-4.pdf:
        - Introduction and Legal Authority
        - Summary and Scope of Proposed Regulations
        - Data Collection Activities
        - Industry Profiles
        - http://epa.gov/ost/guide/cafo/pdf/DDChapters5-7.pdf:
        - Industry Subcategorization for Effluent Limitations Guidelines and Standards
        - Wastewater Characterization and Manure Characteristics
        - Pollutants of Interest
        - http://epa.gov/ost/guide/cafo/pdf/DDChapters8.pdf:
        - Treatment Technologies and Best Management Practices
        - http://epa.gov/ost/guide/cafo/pdf/DDChapters9-11.pdf:
        - NPDES Regulatory Options
        - Technology Options Considered
        - Model Farms and Costs of Technology Bases for Regulation
        - http://epa.gov/ost/guide/cafo/pdf/DDChapters12-14.pdf:
        - Pollutant Reduction Estimates
        - Non-Water Quality Impacts
        - Glossary

- Cost Methodology for the Final Revisions to the National Pollutant Discharge Elimination System Regulation and the Effluent Guidelines for Concentrated Animal Feeding Operations
  Supporting document for the CAFO Final Rule. Provides the data, methodology, and farm level costs for various regulatory options for CAFOs (EPA-821-R-03-004)

    - Part 1 (http://www.epa.gov/npdes/pubs/cafo_cost_method_p1.pdf)
    - Part 2 (http://www.epa.gov/npdes/pubs/cafo_cost_method_p2.pdf)

- Part 3 (http://www.epa.gov/npdes/pubs/cafo_cost_method_p3.pdf)
- Part 4 (http://www.epa.gov/npdes/pubs/cafo_cost_method_p4.pdf)

- Environmental and Economic Benefit Analysis of Final Revisions to the National Pollutant Discharge Elimination System (NPDES) Regulation and the Effluent Guidelines for Concentrated Animal Feeding Operations (CAFOs) Supporting document for the CAFO Final Rule. Environmental and Economic Benefit Analysis (Benefits support document) presents the analysis that the USEPA used to calculate benefits of the revise regulations for concentrated animal feeding operations (CAFOs) (EPA-821-R-03-003)

  - Part 1 (http://www.epa.gov/npdes/pubs/cafo_benefit_p1.pdf)
  - Part 2 (http://www.epa.gov/npdes/pubs/cafo_benefit_p2.pdf)
  - Part 3 (http://www.epa.gov/npdes/pubs/cafo_benefit_p3.pdf)
  - Part 4 (http://www.epa.gov/npdes/pubs/cafo_benefit_p4.pdf)

- Supporting Statement for the Information Collection Request for the Final NPDES and ELG Regulatory Revisions for Concentrated Animal Feeding Operations (CAFOs)
- (http://www.epa.gov/npdes/pubs/cafo_support_for_icr.pdf).
  Information Collection Request (ICR) for the CAFO Final Rule (EPA ICR Number: 1989.02)
- Estimation of National Economic Benefits Using the National Water Pollution Control Assessment Model to Evaluate Regulatory Options for Concentrated Animal Feeding Operations (CAFOs)
  Supporting document for the CAFO Final Rule. Report describes the national-scale water quality model for simulating the water quality and economic benefits that result from various water pollution control policies (EPA-821-R-03-009)

  - Part 1 (http://www.epa.gov/npdes/pubs/cafo_natl_ecobenefit_p1.pdf)
  - Part 2 (http://www.epa.gov/npdes/pubs/cafo_natl_ecobenefit_p2.pdf)
  - Part 3 (http://www.epa.gov/npdes/pubs/cafo_natl_ecobenefit_p3.pdf)
  - Part 4 (http://www.epa.gov/npdes/pubs/cafo_natl_ecobenefit_p4.pdf)

- Non-Water Quality Impact Estimates for Animal Feeding Operations (http://www.epa.gov/npdes/pubs/cafo_nonwaterquality.pdf) Supporting document for the CAFO Final Rule. Report describes the non-water quality impacts analysis for the CAFO Final Rule
- Pollutant Loading Reductions for the Effluent Limitations Guidelines for Concentrated Animal Feeding Operations
  Supporting document for the CAFO Final Rule. Report describes the application of the Groundwater Loading Effects of Agricultural Management Systems (GLEAMS) to determine the effects of various land application controls. The report also describes the methodologies used to calculate production area loads (EPA-821-R-03-007)

- Part 1 (http://www.epa.gov/npdes/pubs/cafo_pollutant_loading_p1.pdf)
- Part 2 (http://www.epa.gov/npdes/pubs/cafo_pollutant_loading_p2.pdf)
- Part 3 (http://www.epa.gov/npdes/pubs/cafo_pollutant_loading_p3.pdf)

- Benefits of Reducing Nitrate Contamination in Private Domestic Wells Under CAFO Regulatory Options (http://www.epa.gov/npdes/pubs/cafo_benefit_nitrate.pdf) Supporting document for the CAFO Final Rule. Report provides estimated reductions in nitrogen and other animal waste-related contaminants that leach through soil and reach ground water (EPA-821-R-03-008)

USDA:

- USDA-NRCS/USEPA Animal Feeding Operation Strategy (http://cfpub.epa.gov/npdes/afo/ustrategy.cfm)
- USDA 2002 Census of Agriculture (USDA 2004) – Provides data on the number of farms for different animal types, as well as estimated production and sales, for each county and state in the US. Written report with summary of state-level data available at www.nass.usda.gov/census/census02/volume1/us/USVolume104.pdf. Searchable database for state- and county-level data is available at www.nass. usda.gov/Census_of_Agriculture or http://agcensus.mannlib.cornell.edu/
- USDA mapping - An on-line interactive map is available that draws from USDA Census of Agriculture database and can be used to identify geographic areas with various amounts of animals or animal farms http://www.nass.usda.gov/ Census_of_Agriculture/SVG/index.asp
- NRCS nutrient and pest management info (http://www.nrcs.usda.gov/technical/ nutrient.html): nutrient management tools and info
- USDA-NRCS agricultural waste management handbook (http://www.wcc.nrcs. usda.gov/awm/awmfh.html): Large guidance document (Table of Contents is 41 pages long). Agricultural Waste Management Field Handbook. National Engineering Handbook (NEH). Part 651. First published in April 1992; last updated June 1, 1999
- NRCS: Animal Feeding Operations (http://www.nrcs.usda.gov/programs/afo/): 2-page website with links
- NRCS: Animal Waste Management (http://www.wcc.nrcs.usda.gov/awm/): nutrient management tools and info
- NRCS National Handbook of Conservation Practices (http://www.nrcs.usda.gov/ technical/Standards/nhcp.html): list of National Conservation Practice Standards
- NRCS Buffer Strips: Common Sense Conservation (http://www.nrcs.usda.gov/ feature/buffers/): info on buffer strips
- NRCS CNMP Technical Guidance (http://www.nrcs.usda.gov/programs/afo/ cnmp_guide_index.html): handbook with guidance for Comprehensive Nutrient Management Planning
- NRCS Economics of Conservation Buffers (http://www.nrcs.usda.gov/technical/ RESS/issues/bufferecon.html): web page with links to info sources on buffer economics
- Costs Associated with Development and Implementation of Comprehensive Nutrient Management Plans (CNMP) for CAFOs (http://www.nrcs.usda.gov/

technical/land/pubs/cnmp1.html): This report details the cost of upgrading facilities and practices on AFOs nationwide to meet CNMP technical guidance, as well the technical assistance needed to plan, design, implement, and follow up on needed structures and practices. Includes links to pdf files for the full report and for individual sections

- ARS National Research Programs and Information (http://www.ars.usda.gov/research/programs.htm): in particular, see Agricultural Watershed Management and Water Quality Protection and Management at http://www.ars.usda.gov/research/programs/programs.htm?npnumber=201&docid=843#agwatman) and also Manure and Byproduct Utilization at http://www.ars.usda.gov/research/programs/programs.htm?NP_CODE=206
- National Program 206: Manure and Byproduct Utilization (http://www.ars.usda.gov/SP2UserFiles/Program/206/WholeNP206AssessmentDocument..): This report by the USDA describes recent research accomplishments in the Manure and Byproduct Utilization National Program. The report is organized into three parts: Nutrients, Emissions and Pathogens containing information on 20 research focus areas
- Water Quality Information Center (http://www.nal.usda.gov/wqic/): USDA Agricultural Research Service's Water Quality Information Center (from the National Agriculture Library). This site provides an extensive list of links to data and other information sources, and one can also search the prototype Database of Online Documents Covering Water and Agriculture (http://grande.nal.usda.gov/wqic/), currently containing 1,442 titles
- Waterborne Pathogens in Agricultural Watersheds - by Barry H. Rosen, R. Croft, E.R. Atwill, S. Wade, and S. Stehman. Natural Resource, Agriculture, and Engineering Service (NRAES) Cooperative Extension, Ithaca, New York, 147 pp., 2001. http://www.nraes.org/publications/nraes147.html
- Confined Animal Production and Manure Nutrients (http://www.ers.usda.gov/publications/aib771/): contains USDA farm census data (Noel Gollehon et al., Agriculture Information Bulletin No. AIB771; 40 pp, June 2001). This publication used census of agriculture data to estimate manure nutrient production and the capacity of cropland and pastureland to assimilate nutrients on a regional and county-by-county basis throughout the US. Water utilities can use this to determine if they are located in a region where nutrient output exceeds input, and thus there may be a reason to be concerned about nutrients from agriculture. Note that the opposite is not a given; i.e., just because there isn't an excess of nutrients does not mean nutrients couldn't be an issue of concern
- Photo gallery for agricultural activities available at http://photogallery.nrcs.usda.gov/. (e.g., see the popdown categories for manure-management and nutrient-management)

USGS:

- AFOs – Effects on Hydrologic Resources and the Environment http://water.usgs.gov/owq/AFO/
- Occurrence of veterinary pharmaceuticals and antibiotics in AFO wastes http://toxics.usgs.gov/regional/emc/animal_feeding.html

- References for new microbial source-tracking and detection techniques (links to other sites and to USGS papers) http://water.usgs.gov/owq/microbial.html
- Trends in Surface Water Quality During Implementation of BMPs in the Mill Creek and Muddy Run Basins, Lancaster, Pennsylvania (http://pa.water.usgs.gov/reports/fs168-99.pdf): Analyses of water samples collected over a 5-year period (1993-1998) during implementation of BMPs indicate statistically significant trends in the nutrient loads to surface waters. USGS Fact Sheet 168-99; January 2000

STATES:

- Links to various state environmental agencies responsible for administering AFO and CAFO regulatory programs: (http://cfpub.epa.gov/npdes/afo/afoinfo.cfm?view=category&link_cat=24)
- Links to description of state AFO and CAFO regulations and requirements (http://cfpub.epa.gov/npdes/afo/afoinfo.cfm?view=category&link_cat=23)
- A list of state environmental agency primary personnel contacts for AFO/NPDES issues: http://cfpub.epa.gov/npdes/contacts.cfm?program_id=7&type=STATE
- A report prepared in 2002 for USEPA provides a compilation of descriptions of state AFO-related regulatory programs and initiatives, with a specific focus on the use of permits or similar mechanisms (Tetra Tech Inc. 2002; http://www.epa.gov/npdes/pubs/statecom.pdf)
- State and Territory Links (http://cfpub.epa.gov/npdes/linkresult.cfm?program_id=7&link_category=2&view=link) - Provides links to state web sites that provide specific information related to the NPDES permit program
- Texas Commission on Environmental Quality (http://www.tceq.state.tx.us/nav/permits/cafo.html): CAFO rule permit guidance
- Nebraska Department of Environmental Quality, Facts Sheets & Guidance Documents (http://www.deq.state.ne.us/) - Once you get to this web site, to access livestock waste topics and forms, click on "NDEQ Programs", then click on the triangle next to "Agriculture Programs", then click on the triangle next to "Livestock Waste Control Program". This shows sub-topics:
- "Livestock Waste Control Program" is a report of the numbers of recent permit applications. At the bottom of the page is a link to current "Livestock Permit Applications on Public Notice"
- "Applications, Forms, and Publications" contains blank forms, guidance documents, and examples of completed forms
- "Title 130" has the entire Nebraska law concerning livestock waste management in individual chapters
- Copy of CAFO Rule Fact Sheet #1 (http://www.iowabeefcenter.org/heartlandwq/Education/Fact1.pdf; from LPES): includes a page of contacts for NE, KS, IA, and MO, and USEPA (October 2003)
- The State of Idaho published on-line their Nonpoint Source Pollution Program Annual Report which includes a listing of a number of projects sponsored or conducted by the state for watershed, AFO, and BMP related projects (for 2004 see http://www.state.sd.us/denr/DFTA/WatershedProtection/NPS/04NPSAnnualReport.pdf)

- Aquaculture systems and waste management in general, and with specifics to the State of Idaho, are described by the Idaho Department of Environmental Quality at http://www.deq.idaho.gov/water/prog_issues/agriculture/aquaculture.cfm
- Waste management systems for aquaculture facilities are discussed in detail by Idaho DEQ (2002, http://www.deq.idaho.gov/water/prog_issues/waste_water/pollutant_trading/aquaculture_guidelines.pdf

COOPERATIVE EXTENSION SERVICES:

- University of Nebraska manure matters website (http://www.ianr.unl.edu/manure): includes workbooks on manure application, storage, and odor management
- North Carolina manure matters (http://www.bae.ncsu.edu/programs/extension/publicat/wqwm/animops.html): a list of publications
- Iowa manure matters (http://www.extension.iastate.edu/Pages/communications/EPC/): a newsletter (*Iowa Manure Matters: Odor and Nutrient Management)* published by Iowa State University Extension (http://www.extension.iastate.edu/), with funding support from the USDA Natural Resources Conservation Service (http://www.ia.nrcs.usda.gov/)
- Ohio State University (http://www.ag.ohio-state.edu/~farmnet/links/nutmgt.html): lots of links to other websites
- Ohio Livestock Manure and Wastewater Management Guide (http://ohioline.osu.edu/b604/.) (OSU bulletin 604)
- Purdue Animal Manure Solutions (http://danpatch.ecn.purdue.edu/~epados/pams/): Purdue University site that has lots of web links and articles on manure management
- Managing Phosphorus from Animal Manure (http://osuextra.com/pdfs/F-2249web.pdf): Fact sheet from the Oklahoma State University – Oklahoma Cooperative Extension Service. Proper management of P sources and P transport over the landscape can reduce negative impacts of P on water quality
- Information and descriptions of BMPs for managing nutrient and pesticide wastes are available at http://www.oznet.ksu.edu/waterquality/welcome.htm.
- Good summaries of information about phosphorus in the agricultural environment is provided at:

  - http://www.nrcs.usda.gov/technical/land/pubs/wp08text.html
  - http://www.utextension.utk.edu/publications/pbfiles/PB1645%20.pdf
  - http://www.ext.vt.edu/pubs/grains/424-029/424-029.html

- Information on managing manure phosphorus is available at

  - http://muextension.missouri.edu/explore/agguides/soils/g09182.htm
  - http://osuextra.com/pdfs/F-2249web.pdf

- Select BMPs for managing phosphorus are described at:

  - http://www.oznet.ksu.edu/library/crpsl2/mf2321.pdf
  - http://www.utextension.utk.edu/publications/pbfiles/PB1645%20.pdf

OTHER SOURCES:

- Agricultural BMP Effectiveness Database (PSU 1992). Developed for USEPA by Pennsylvania State University and others. http://iaspub.epa.gov/edr/edr_proc_qry.navigate?P_LIST_OPTION_CD=CSDIS&P_REG_AUTH_IDENTIFIER=1&P_DATA_IDENTIFIER=90637&P_VERSION=1
- BMP Database for Nutrient Reduction (Yagow et al. 2002). Developed by Virginia Polytechnic Institute and State University for the USEPA Chesapeake Bay Program. http://asae.frymulti.com/abstract.asp?aid=7565&t=2
- American Society of Agricultural and Biological Engineers (ASABE, formerly ASAE) – a searchable technical library for all publications from the ASABE is available at http://asae.frymulti.com/
- Cornell University website with fact sheets, case studies, and technical information for dairy manure management and treatment systems. http://www.manuremanagement.cornell.edu
- The University of California (2005) published a good report on managing dairy manure http://www.waterboards.ca.gov/centralvalley/available_documents/dairies/uc-committee-of-experts-rpt.pdf
- National Association of Conservation Districts maintains a website with links to conservation district websites in each state. http://www.nacdnet.org/resources/cdsonweb.html
- National Organization of State Conservation Agencies (NOSCA) – list if member agencies. http://nascanet.org/agencies.htm
- Center for Watershed Protection - studies of watershed planning and protection, and BMP effectiveness http://www.cwp.org
- Manure Management Planner Home Page - a Windows-based computer program used to create manure management plans for crop and animal feeding operations. This planner currently supports 31 states by automatically generating fertilizer recommendations and estimating manure N availability based on each state's Extension or NRCS guidelines. http://www.agry.purdue.edu/mmp/
- University of California at Davis, Groundwater Cooperative Extension Program, Dairy Nutrient Management and Groundwater Quality research (http://groundwater.ucdavis.edu/gw_201.htm). A list of links to websites with various resources for dairy AFO operations is available at http://groundwater.ucdavis.edu/dairy_links.htm
- National Farmstead Assessment System (http://www.wisc.edu/farmasyst/): Farm*A*Syst
- Natural Resource, Agriculture, and Engineering Service (http://www.nraes.org/): NRAES is the Natural Resource, Agriculture, and Engineering Service (formerly known as the Northeast Regional Agricultural Engineering Service). NRAES is an interdisciplinary, issue-oriented program sponsored by cooperative extension of fourteen member land grant universities. The program is guided by the NRAES Committee (http://www.nraes.org/ncommittee.html), which includes a representative from each member university. NRAES is a hub for the integration of research, teaching, and extension and for multi-state activity. The mission of NRAES is to assist faculty and staff at member universities in increasing the public availability of research- and experience-based knowledge

- Conservation Technology Information Center (http://www.ctic.purdue.edu/CTIC/CTIC.html): Some information is available at the Core 4 Conservation page (http://www.ctic.purdue.edu/Core4/Core4Main.html)
- Midwest Plan Service (http://www.mwpshq.org/): Includes a listing of "affordable, practical" publications for agriculture
- Best Management Practices: Land Application of Animal Manure (http://ohioline.osu.edu/agf-fact/0208.html.): 10-page report
- Liquid Manure Application Methods (http://www.ext.colostate.edu/pubs/livestk/01223.html): 5-page report from Colorado State University Cooperative Extension
- Pathogen Movement and Survival in Catchments, Groundwaters and Raw Water Storages. CRC summary of research and a series of fact sheets are available describing the factors affecting pathogen transport in the environment and engineered systems (CRC 2004). http://www.waterquality.crc.org.au/dwfacts/techfact_pathogen_movement.PDF
- Livestock and Poultry Environmental Stewardship (LPES) Curriculum (http://www.lpes.org/): Includes 24 CAFO Rule fact sheets (http://www.lpes.org/CAFO.html) that are very helpful. The CAFO Fact Sheets are a collection of 24 publications that address questions that educators and producers are most likely to have about what the CAFO regulations mean, how they will affect livestock and poultry production facilities, and what producers must do to comply with the regulations

General Information (http://www.lpes.org/CAFO_PDF_Gen.html):

- LPES CAFO Fact Sheet #1: A Brief History and Background of the EPA CAFO Rule
- LPES CAFO Fact Sheet #2: Do I Need an NPDES Permit for My Livestock and Poultry Operation?
- LPES CAFO Fact Sheet #3: How Soon Must I Apply for an NPDES Permit?
- LPES CAFO Fact Sheet #4: How Do I Apply for an NPDES Permit?
- LPES CAFO Fact Sheet #5: What if My Operation is an AFO But Not a CAFO?
- LPES CAFO Fact Sheet #6: What CAFO Reports Must I Submit? (includes Annual Report Record Sheet)
- LPES CAFO Fact Sheet #7: Alternate Treatment Systems

Production Area (http://www.lpes.org/CAFO_PDF_Prod.html):

- LPES CAFO Fact Sheet #10: CAFO Requirements for the Production Area
- LPES CAFO Fact Sheet #11: CAFO Requirements for Dairy Barnyards
- LPES CAFO Fact Sheet #12: CAFO Requirements for the Beef Production Area
- LPES CAFO Fact Sheet #13A: CAFO Requirements for the Poultry Production Area
- LPES CAFO Fact Sheet #13B: CAFO Requirements for Large Swine Operations
- LPES CAFO Fact Sheet #13C: CAFO Requirements for Veal Operations

- LPES CAFO Fact Sheet #14: Pasture-based Operations: Could Part of My Operation Be an AFO?
- LPES CAFO Fact Sheet #15: Liquid Level Markers for Uncovered Manure Storages and Lagoons

Land Application (http://www.lpes.org/CAFO_PDF_Land.html):

- LPES CAFO Fact Sheet #20: What is Required in a Nutrient Management Plan?
- LPES CAFO Fact Sheet #21: What is the Agricultural Storm Water Exemption?
- LPES CAFO Fact Sheet #22: Land Application Setback and Buffer Requirements for NPDES Permitted Large CAFOs
- LPES CAFO Fact Sheet #23: Transfer of Manure or Wastewater Off Farm
- LPES CAFO Fact Sheet #24: Technology Options to Comply with Land Application Rules
- LPES CAFO Fact Sheet #25: Making Decisions about Application Rates
- LPES CAFO Fact Sheet #26: What Records Must I Maintain for Land Application?

Farm Bill and Financial Assistance (http://www.lpes.org/CAFO_PDF_Farm.html):

- LPES CAFO Fact Sheet #30: Financial and Technical Assistance Available to CAFO Owners/Operators
- LPES CAFO Fact Sheet #31: EQIP Financial and Technical Assistance Available to CAFO Owners and Operators

- Oklahoma Litter Market – information source for poultry litter issues http:// /biosystems2.okstate.edu/poultry/marketing_incentives.asp
- National Dairy Environmental Stewardship Council provides information on cost-effective dairy manure management practices (http://www.suscon.org/ dairies/ ndesc.asp), including a report (NDESC 2005) available at http://www.suscon.org/ dairies/pdfs/COST%20EFFECTIVE%20AND%20ENVIRONMENTALLY.pdf. The goal of the NDESC is to identify successful treatment options for dairy waste that are both environmentally beneficial and economically viable, and to recommend strategies for increased adoption of these practices
- Fact sheets on poultry litter - http://fp.okstate.edu/zhailin/extension.htm
- National Pork Board (http://www.porkboard.org/cafo.asp): Pork economics and markets, link to 15 CAFO Rule fact sheets (from LPES), other pork info
- Farmed Animal Watch (http://www.farmedanimal.net/faw/faw5-7.htm): Includes miscellaneous information about farmed animal operations. Seems to focus on poultry operations. Their newsletters contain some marketing/production info
- US Poultry & Egg Association (http://www.poultryegg.org/). Links include Communications, Environment, Food Safety, Fowl Processor Council, International Poultry Scientific Forum, Membership, Poultry Protein & Fat Council, Poultry Water Quality Consortium, Research, Seminars/Education
- The Aquaculture Network Information Center (AquaNIC; http://aquanic.org/) was conceived in 1994 by the former USDA-Extension Service (currently Cooperative State Research, Education and Extension Service; http://www.csrees.usda.gov/) to

serve as a gateway to the world's electronic aquaculture resources. Includes mostly links, even to fish recipes

- The Southern Regional Aquaculture Center (SRAC, http://www.msstate.edu/dept/srac/) provides a series of fact sheets about aquaculture and aquaculture waste management.
- Waste management systems for aquaculture facilities, including BMPs to reduce aquaculture waste, are summarized by Miller and Semmens (2002; http://www.wvu.edu/~afexten/aquaculture/waste02.pdf.)
- Pollution Scorecard (http://www.scorecard.org/): Sponsored by the Environmental Defense Fund (http://www.environmentaldefense.org/home.cfm), an environmental advocate group, this website includes information on various pollution sources throughout the US. One part of the site (http://www.scorecard.org/env-releases.aw/.) includes maps and other information that compares the amount of animal waste produces in different states or counties. Also includes a state map of North Carolina showing where the hog operations are located (http://www.scorecard.org/env-releases/aw/nc-hog-campaign.tcl)
- Sierra Club (http://www.sierraclub.org/factoryfarms/resources/factory_farm_studies.doc): The Sierra Club has compiled a list of studies done a variety of institutions on comprehensive farm studies. This document presents a brief summary of these studies along with citations. http://www.sierraclub.org/factoryfarms/resources/factory_farm_studies.doc
- Effects of Voluntary Agricultural Best Management Practice Implementation on Water Quality (http://www.engr.uga.edu/service/extension/agp2/resources/publication/A-G/Effects%20of%20Voluntary%20A18.doc): This literature review from researchers at The University of Georgia was conducted to see if there is evidence that voluntary programs, specifically related to livestock production and nutrient management, are effective. This report lists and summarizes 21 studies that supplied some indication of the effectiveness of voluntary programs and a summary to provide some interpretations of the results
- Costs and Environmental Effectiveness for Nutrients BMPs for Dairy Farms in the Lake Fork Reservoir Watershed (http://tiaer6.tarleton.edu/pdf/PR9807.pdf): October 1998 report from researchers at Tarleton State University, Stephenville, Texas

AwwaRF REPORTS:

- Impacts of Major Point and Non-Point Sources on Raw Water Treatability (AwwaRF #2616; http://www.awwarf.org/research/TopicsAndProjects/projectSnapshot.aspx?pn=2616). Evaluates the impacts of point and non-point source discharges on drinking water source water quality, drinking water treatability, and water treatment costs. Also evaluates the potential for using best available technologies and best management practices to mitigate impacts prior to the water treatment plant intake. Published in 2003
- Source Water Protection Alliances Between Water Utilities and Agricultural Operations. (AwwaRF #2781; http://www.awwarf.org/research/TopicsAndProjects/projectSnapshot.aspx?pn=2781) The research project was designed to provide drinking water utilities with guidance on how they can build successful source water protection alliances with farmers and agricultural organizations to

promote agricultural practices that minimize runoff and protect the quality of drinking water sources. A webpage was developed. Published in 2005

- Source Water Protection Reference Manual (CD-rom only; AwwaRF #2651; http://www.awwarf.org/research/TopicsAndProjects/projectSnapshot.aspx?pn=2651) This valuable reference for water utilities comprises information on four main facets of watershed protection: best management practices, planning, regulations, and water supplier experience. The frequently-asked-questions section and search engine make it easy to use. Updates and expands AwwaRF Project #903 to include changes in the SDWA amendments, new advances in watershed management, and Canadian and UK initiatives. Includes a planning section, Web hot links, and comparison of BMP effectiveness. *Available only in CD-ROM format.* Published in 2002
- Guidance to Utilities on Building Alliances With Watershed Stakeholders. (AwwaRF #468; http://www.awwarf.org/research/TopicsAndProjects/projectSnapshot.aspx?pn=468) Identifies typical watershed stakeholders and their objectives in basin planning. Also, develops procedures for building win-win alliances between water utilities and stakeholders for the purpose of overcoming constraints to planning, managing, and developing long-term sustainable drinking water supplies (Raucher and Goldstein 2001)

APPENDIX B

UTILITY SURVEY

Utility Survey (AwwaRF Project #3020)

Concentrated Animal Feeding Operations (CAFO): Water Utility Issues and Regulatory Controls

Please provide the requested information about animal feeding operations (AFO) or concentrated animal feeding operations (CAFO) in your watershed and return electronic or hardcopy of this survey to Ms. Meredith Sherrill via email, mail, or FAX as outlined below. It is expected that most people responding to this survey will provide fairly short responses in the limited space provided in this survey, but please feel free to attach additional sheets if you would like to provide additional information. Please contact Dr. Richard Gullick if you have any questions.

Contact for questions:

Richard W. Gullick, Ph.D.
Environmental Engineering & Technology, Inc.
73 N. Grove St., Berlin, NJ 08009
Phone (856) 875-6140, Fax (856) 875-6142
e-mail: rgullick@eetinc.com

Please return completed survey by September 30, 2005 to:

Mr. Richard Brown
Environmental Engineering & Technology, Inc.
712 Gum Rock Court, Newport News, VA 23606
Phone (757) 873-1534, Fax (757) 873-2392
e-mail: rbrown@eetinc.com

1. Provide utility contact information and describe source water, treatment, and other characteristics

1.1. Contact information

Utility Name:	Contact Name:
City/State:	Title:
Phone:	FAX:
Email:	

1.2. Do you have AFOs or CAFOs in your watershed?

Yes No (and we have investigated the possibility) Don't know

Note: **If answer is No or Don't Know, please do not complete the rest of the survey (but please do let us know and provide the contact information above in Question 1.1). If answer is Yes, please complete the rest of the survey.**

1.3. Source water and treatment description

	Treatment Facility		
	#1	#2	#3
PWSID(s)			
Facility Name			
City, State			
Capacity (mgd)			
Current Annual Average Production (mgd)			
Population served (not connections)			
Source water type: GW, river, lake/ reservoir, GWUDI, purchased, etc. (include name of lake, river, aquifer, or other source as appropriate)			
Briefly describe treatment process: e.g., alum coagulation, clarification, dual-media filtration, free chlorine disinfection			

2. **AFO/CAFO impacts in watershed**

2.1. **What types of AFOs or CAFOs are present in your watershed? Please complete each box with "yes", "no", or "don't know", and identify others not on this list in the space provided below**

No AFO/CAFO in watershed	Swine	Cattle - Beef	Cattle - Dairy	Poultry – chicken	Poultry – turkey	Aquaculture

Other (please list):

____

2.2. **Which is generally considered the most important type of AFO/CAFO in your watershed? (in each box, identify one of the AFO/CAFO types from previous question)**

Most AFO/CAFO sites in our watershed	
The largest AFO/CAFOs in our watershed, in terms of waste production	
The most economically important AFO/CAFOs to residents in our watershed	
Most chronic (i.e., day-to-day) releases of contaminants or nutrients of concern to our source water are probably from this type AFO/CAFO	
Greatest risk for acute (i.e., large, one-time) releases of contaminants or nutrients of concern to our source water are probably from this type AFO/CAFO	

If the answer to these questions is not the same type of AFO/CAFO, please explain below or on a separate sheet of paper

____

2.3. **Are wastes from AFOs or CAFOs currently impacting source water quality for your utility, or have they impacted water quality in the past? (check one)**

Yes, we have direct evidence	No direct evidence but suspect it is true	There are AFO/CAFO in watershed but no known or suspected impacts	Don't know	No AFO/CAFO in our watershed

2.4. **Please identify name and contact information for state or local extension agent, USDA contact, or other person you would recommend we contact regarding AFO/CAFO impacts in your watershed, particularly those your utility have found particularly useful or knowledgeable.**

Please list below the organization, contact name, position/title, phone, and email address:

____

2.5. **Who is the primary contact or who has primary responsibility for watershed issues at your utility?**

Please list name, position/title, phone, and email address:

____

3. **Treatment for control of contaminants originating from AFO/CAFO wastes**

3.1. **Which contaminants are of concern for your utility, both now and in the future? Do you consider AFOs or CAFOs significant sources in your watershed? Please complete all boxes with either "yes", "no", "DN" (don't know), or "NA" (not applicable).**

Contaminants	**Current concern**	**Possible future concern**	**AFO/CAFOs may be significant sources in our watershed**	**Sources other than AFO/CAFOs are probably the predominant sources of these contaminants in our watershed**
Cryptosporidium and *Giardia*				
Other microbial contaminants (e.g., virus, bacteria)				
Hormones, antibiotics, and other pharmaceuticals				
Precursors of currently regulated DBPs				
Precursors of DBPs not currently regulated (e.g., precursors of NDMA and other nitrogenous DBPs)				
Insecticides and herbicides				
Other organic contaminants not listed above				
Heavy metals				
Nutrients (nitrogen and phosphorus)				
Other contaminants (please list)				

3.2. **Have your treatment operations been modified in response to measured or suspected contaminants from AFO/CAFO wastes?**

YES / NO **If yes, please briefly describe:**

3.3. **Are you contemplating or developing modifications to treatment infrastructure or operational practices specifically due to potential contaminants from AFO/CAFO wastes?**

YES / NO **If yes, please briefly describe:**

---

4. **Current and future source water protection (SWP) efforts**

4.1. **Did the source water assessment (SWAP) for your intake(s) identify AFOs or CAFOs as potential contaminant sources?** YES / NO

4.2. Do you have a source water protection program? YES / NO

If not, do you currently have plans to develop one? YES / NO

4.3. Have you established a dialogue with AFO/CAFO owners and operators in your watershed, and/or with state agencies overseeing AFO/CAFO facilities in your state?

YES / NO **If yes, please briefly describe.**

4.4. Please list and briefly describe SWP measures your utility (or watershed partners, including AFOs/CAFOs) has implemented relative to AFO/CAFO wastes. What are your measures of success and are you achieving your objectives? Please attach additional information if available.

4.5. What are your primary future goals for source water protection, especially in relation to AFO/CAFO wastes?

4.6. What would you most like for your utility to get out of this AwwaRF study? What future studies regarding source water protection, in particular related to contaminants from AFO/CAFO sources, would you like to see addressed in this or future AwwaRF studies?

4.7. May we follow up with you to obtain additional information regarding CAFOs in your watershed?

YES / NO **If yes, whom may we contact (name, position/title, phone, and e-mail address)?**

REFERENCES

Agricultural Animal Waste Task Force, Nicholas School of the Environment Duke University. 1996. Policy Recommendations for Management of Agricultural Animal Waste in North Carolina. Durham, NC: Duke University. Available: <http://lead.virtualcenter.org/en/dec/toolbox/Indust/agwaste.doc>

Amirtharajah, A., M.H. Young, K.D. Pennell, J.L. Steiner, D.S. Fisher, D.M. Endale, and J.P. Campbell. 2002. Field Transport of *Cryptosporidium* Surrogate in a Grazed Catchment. Denver, Colo.: AwwaRF.

APHA, AWWA, and WEF (American Public Health Association, American Water Works Association, and Water Environment Federation). 2005. *Standard Methods for the Examination of Water and Wastewater*: 21st ed. Washington, D.C.: APHA.

ASAE (American Society of Agricultural Engineers). 2005. *Standard: Manure Production and Characteristics* ASAE D384.2; Final version May 2005. St. Joseph, Mich.: ASAE.

Atherholt, Thomas B., Mark W. LeChevallier, William B. Norton, and Jeffrey S. Rosen. 1998. Effect of Rainfall on Giardia and Crypto. *Jour. AWWA,* 90(9):66-80.

Atwill, E. R., L. Hou, B.M. Karle, T. Harter, K.W. Tate, and R.A. Dahlgren. (2002). Transport of *Cryptosporidium* parvum Oocysts through Vegetated Buffer Strips and Estimated Filtration Efficiency. *Appl. Environ. Microbiol,* 68(11):5517-5527. Available: <http://aem.asm.org/cgi/content/full/68/11/5517>

Auvermann, Brent W. 2005. Beef Cattle Feedyards. Presented at the Workshop on Concentrated Animal Feeding Operations (CAFOs): Water Utility Issues and Regulatory Controls (AwwaRF Project #3020 meeting), Cleveland, Ohio, September 19-20, 2005.

AWWA (American Water Works Association). 2004. *Formal Comments of the American Water Works Association on the Long-Term 2 Enhanced Surface Water Treatment Proposed Rule, Appendices 1 and 2.* Water Docket OW-2002-0039. Washington, DC: AWWA. Available: <http://www.awwa.org/Advocacy/govtaff/documents/AWWACommentsonLT2ESWTRJan92004.pdf>

AWWA (American Water Works Association). 1999. *Source Water Protection: Effective Tools and Techniques You Can Use: 1999 Participant Manual.* Denver, Colo.: AWWA.

AWWA Microbiological Contaminants Research Committee. 1999a. Committee Report: Emerging Pathogens – Bacteria. *Jour. AWWA*, 91(9):101-109.

AWWA Microbiological Contaminants Research Committee. 1999b. Committee Report: Emerging Pathogens – Viruses, Protozoa, and Algal Toxins. *Jour. AWWA*, 91(9)110-121.

Ballester, Nicola A., and James P. Malley. 2004. Sequential Disinfection of Adenovirus Type 2 with UV-Chlorine-Chloramine. *Jour. AWWA*, 96(10):97-103.

Baticados, M.C.L., and J.O. Paclibare. 1992. The Use of Chemotherapeutic Agents in Aquaculture in the Philippines. In *Diseases in Asian Aquaculture I.* Edited by M. Shariff, R.P. Subasinghe, and J.P. Arthur. Manila, Philippines: Fish Health Section, Asian Fishery Society.

Beeman, Perry. 2002. Nitrate Levels Surge in D.M. [Online]. Des Moines Register.com, May 17, 2002. Available: <http://desmoinesregister.com/news/stories/c5903220/18215372.html>. [cited July 15, 2006]

Bertolucci, G., G. Gilli, E. Carraro, D. Giacosa, and M. Puppo. 1998. Influence of Raw Water Storage on *Giardia*, *Cryptosporidium* and Nematodes. *Water Science and Technology*, 37(2):261-267.

Bitton, Gabriel. 2005. *Microbial Indicators of Fecal Contamination: Application to Microbial Source Tracking*. Gainesville, Fl.: University of Florida. Available: <http://www.florida-stormwater.org/pdfs/FSAMicrobialSourceTrackingReport.pdf>

Carter, Thomas A., and Matt Poore. 1995. Deep Stacking Broiler Litter as a Feed for Beef Cattle. Publication Number AG-515-2. North Carolina Cooperative Extension Service, North Carolina State University, Raleigh, N.C.

CCME (Canadian Council of Ministers of the Environment, Federal-Provincial-Territorial Committee on Drinking Water and the CCME Water Quality Task Group). 2002. *From Source to Tap: Guidance on the Multi-Barrier Approach to Safe Drinking Water*. Winnipeg, Manitoba, Canada: CCME. Available: <http://www.ccme.ca/assets/pdf/mba_guidance_doc_e.pdf>

CDM (Camp Dresser & McKee, Inc.). 2002. Source Water Protection Reference Manual [CD-ROM]. Available: AwwaRF and AWWA, Denver, Colo.

Chaney, R. et al. 2003. Heavy Metal Contaminants in Composts and Manures. Presented at University of Wisconsin-Madison.

City of Mankato. 2005. Drinking Water Report. City of Mankato, Minnesota. Available: <http://www.ci.mankato.mn.us/utility/WaterReport.pdf>

City of Tulsa and INCOG (Indian Nations Council of Governments). 1999. Watershed Restoration Action Strategy (WRAS) for the Eucha/Spavinaw Watershed. INCOG, Tulsa, Okla. Available: <http://www.deq.state.ok.us/WQDnew/pubs/eucha_watershed_wras_final.pdf>

CRC (Cooperative Research Centre for Water Quality and Treatment). 2004. Pathogen Movement and Survival in Catchments, Groundwaters and Raw Water Storages. Salisbury, South Australia: CRC. Available: <http://www.waterquality.crc.org.au/dwfacts/techfact_pathogen_%20movement.PDF>

CSREES (USDA Cooperative State Research, Education, and Extension Service). 2005. Nutrient Management Decision and Education Support System [Online]. Available <http://cris.csrees.usda.gov/cgi-bin/starfinder/0?path=fastlink1.txt&id=anon&pass=&search=AN=0192683&format=WEBFMT7>. [cited July 19, 2006]

Curriero, F., J. Patz, J. Rose, and S. Lele. 2001. The Association Between Extreme Precipitation and Waterborne Disease Outbreaks in the United States, 1948-1994. American Journal of Public Health, 91(8):1194-1199.

Darnault, Christophe J.G. 2000. Ph.D. diss. Cornell University, Ithaca, N.Y.

Darnault, Christophe J.G., Patricia Garnier, Young-Jin Kim, Kristina L. Oveson, Tammo S. Steenhuis, J.-Yves Parlange, Michael Jenkins, William C. Ghiorse, and Philippe Baveye. 2003. Preferential Transport of *Cryptosporidium parvum* Oocysts in Variably saturated Subsurface Environments. *Water Environment Research*, 75(2):113-120. <http://www.ars.usda.gov/SP2UserFiles/Place/66120900/ManureManagementToProtectTheEnvironment/2003mbjJ09.pdf>

Darnault, Christophe J.G., Tammo S. Steenhuis, Patricia Garnier, Young-Jin Kim, Michael B. Jenkins, William C. Ghiorse, Philippe Baveye, and J.-Yves Parlange. 2004. Preferential Flow and Transport of *Cryptosporidium parvum* Oocysts through the Vadose Zone: Experiments and Modeling. *Vadose Zone Journal*, 3:262-270.

Davies, Cheryl, Christine Kaucner, Nanda Altavilla, Nicholas Ashbolt, Christobel Ferguson, Martin Krogh, Wim Hijnen, Gertjan Medema, and Daniel Deere. 2005. *Fate and Transport of Surface Water Pathogens in Watersheds*. Denver, Colo.: AwwaRF.

Deininger, R.A. and J-Y. Lee. 1998. "Rapid Determination of Bacteria in Water." In: *Proc. of the 1998 AWWA Water Quality Technology Conference, San Diego, CA*. Denver, CO: AWWA.

DeWolfe, James R. 2006. *Water Residuals to Reduce Soil Phosphorus*. Denver, Colo.: AwwaRF.

DeWolfe, James R. 2003. Use of Water Treatment Residuals to Bind Phosphorus to Protect Surface Water Quality. In *Proc. of the AWWA Water Quality Technology Conference*. Philadelphia, Pa.: AWWA.

Eccles, J.P., E.J. Fricker, and C.R. Fricker. 1998. Rapid Detection of Coliforms and Fecal Coliforms Within 8 hours Using the Colifast CA-100 Analyzer and Media. In *Proc. of the 1998 AWWA Water Quality Technology Conference*. Denver, Colo.: AWWA.

Edberg, Stephen C., Frances Ludwig, Darrell B. Smith, and Martin J. Allen. 1991. *The Colilert System for Total Coliforms and Escherichia coli*. Denver, Colo.: AwwaRF.

Edberg, Stephen C., and Darrell B. Smith. 1994. *Comparison of the Colilert Method and Standard Fecal Coliform Methods*. Denver, Colo.: AwwaRF.

EE&T (Environmental Engineering & Technology, Inc.). 2005. *Quality Assurance Project Plan* (Project #3020) *Concentrated Animal Feeding Operations (CAFOs): Water Utility Issues and Regulatory Control*. Newport News, Va.: EE&T, Inc.

Ernst, C. 2004. *Protecting the Source: Land Conservation and the Future of America's Drinking Water*. Washington, D.C.: Trust for Public Land and Denver, Colo.: AWWA.

Ernst, C., and K. Hart. 2005. *Path to Protection: Ten Strategies for Successful Source Water Protection*. Washington, D.C.: Trust for Public Land.

Ferguson, Christobel, Nanda Altavilla, Nicholas J. Ashbolt, and Daniel Deere. 2003. Prioritizing Watershed Pathogen Research. *Jour. AWWA*, 95(2):92-102.

Finstein, Melvin S. 2004. Protecting Watersheds from *Cryptosporidium* in Manure. *Jour. AWWA*, 96(2)114-116.

Fletcher, Angie, Susan Davis, and Grantley Pyke. 2004. *Water Utility/Agricultural Alliances: Working Together for Cleaner Water*. Denver, Colo.: AwwaRF.

Frey, Michelle and Laurie Sullivan. 2004. *Practical Application of Online Monitoring*. Denver, Colo.: AwwaRF.

Gallagher, K., J. Leech, and H. Stowe. 1992. Protein, energy and dry matter consumption by racing thoroughbreds: A field survey. *J. Equine Vet Sci.*, 12:43–48.

Garbarino, J.R., A.J. Bednar, D.W. Rutherford, R.S. Beyer, and R.L. Wershaw. 2003. Environmental Fate of Roxarsone in Poultry Litter. I. Degradation of Roxarsone during Composting. *Environ. Sci. & Technol.*, 37(8):1509-1514.

General Assembly of North Carolina. 1997. Session Law 1997-458, *House Bill 515*. An Act To Enact The Clean Water Responsibility and Environmentally Sound Policy Act, A Comprehensive and Balanced Program To Protect Water Quality, Public Health, and The Environment. Available: <http://www.ncga.state.nc.us/Sessions/1997/Bills/House/HTML/H515v6.html>

General Assembly of North Carolina. 2003. Session Law 2003-266, *Senate Bill 593*. An Act To Extend The Moratoria On Construction Or Expansion Of Swine Farms. Available: <http://www.ncga.state.nc.us/EnactedLegislation/SessionLaws/PDF/2003-2004/SL2003-266.pdf>

GeoLogics Corporation (Alexandria, VA). 2004. EPA Regional Priority AFO Science Question Synthesis Document - Pharmaceuticals and Pathogens. Workshop Review Draft: Supporting Documentation for the EPA Regional Science Workshop on Animal Feeding Operations (AFOs) - Science and Technical Support Needs, December 6-9, 2004, College Park, Md. Washington, D.C.: Office of Science Policy and Office of Research and Development. Available: <http://www.manure.umn.edu/regulatory/4Pharmaceuticals %20and%20Pathogens.pdf>

Gollnitz, William D. Jennifer L. Clancy, J. Brock McEwen, and Stephen C. Garner. 2005. Riverbank Filtration for IESWTR Compliance. *Jour. AWWA*, 97(12):64-76.

Government of Alberta (Canada). 2004. Manure Composting Manual. Available <www1.agric. gov.ab.ca/$department/deptdocs.nsf/all/agdex8875?opendocument>

Grayman, Walter M., Rolf A. Deininger, and Richard M. Males. 2001. *Design of Early Warning and Predictive Source-Water Monitoring Systems*. Denver, Colo.: AwwaRF.

Grayman, Walter M., Rolf A. Deininger, Richard M. Males, and Richard W. Gullick. 2004. *Source Water Early Warning Systems*. In *Water Supply Systems Security*. Edited by Larry W. Mays. New York, N.Y: McGraw-Hill.

Gullick, Richard W. 2003 *Monitoring Systems for Early Warning of Source Water Contamination*. Voorhees, N.J.: American Water Works Service Company, Inc.

Gullick, Richard W., Walter M. Grayman, Rolf A. Deininger, and Richard M. Males. 2003. Design of Early Warning Monitoring Systems for Source Waters. *Jour. AWWA*, 95(11): 58-72.

Hannawald, James E. 1999. Alternative Waste Management Technologies: Summary of Available Resources. USDA-NRCS. Available: <http://www.nrcs.usda.gov/technical/ ECS/nutrient/manureutil.doc>

Hanselman, Travis A., Donald A. Graetz, and C. Wilkie. 2003. Manure-Borne Estrogens as Potential Environmental contaminants: A Review. *Environ. Sci. & Technol.*, 37(24)5471-5478.

Hargesheimer, Erika, Osvaldo Conio, and Jarka Popovicova 2002. Online Monitoring for Drinking Water Utilities. Denver, Colo.: AwwaRF.

Harner, J. P., and P. K. Kalita. 1999. Vegetative Filters for Improving Environmental Quality. Cattlemen's Day 1999. Available: <http://www.oznet.ksu.edu/library/lvstk2/sections/ SRP831s/831_40.pdf>

Harter, Thomas H. 2006. Personal communication.

Harter, Thomas, H. Davis, Marsha C. Mathews, and Roland D. Meyer. 2002. Shallow Groundwater Quality on Dairy Farms with Irrigated Forage Crops. *Jour. of Contaminant Hydrology*, 55(3-4):287-315.

Harter, Thomas H., Roland D. Meyer, and Marsha C. Mathews. 2002. Nonpoint source pollution from animal farming in semi-arid regions: Spatio-temporal variability and groundwater monitoring strategies. In: Future Groundwater Resources at Risk, Edited by L. Ribeiro, 2002. In *Proc. of the 3rd International Conference, Lisbon, Portugal, June 2001*; p. 363-372.

Harter, Thomas, Marsha Campbell-Matthews, and Roland Meyer. 2005. Project: Dairy Nutrient Management and Groundwater Quality. [Online]. University of California at Davis, Groundwater Extension Service. Davis, Calif. <http://groundwater.ucdavis.edu/gw_ 201.htm>

Herpel, Rachael. 2004. Source Water Assessment & Protection Workshop Guide. 2nd ed. Lincoln, Neb.: Groundwater Foundation.

Hopper, K. and C. Ernst. 2005. Source Protection Handbook: Using Conservation to Protect Drinking Water Supplies. Washington, D.C.: Trust for Public Land and Denver, Colo.: AWWA.

Humenik, Frank, Joe Zublena, Jim Barker, and Tara Disy. 1996. Constructed Wetlands for Animal Wastewater Treatment. North Carolina Cooperative Extension Service, Publication AG-473-13. Available: <http://www.bae.ncsu.edu/programs/extension/publicat/wqwm/ag473_13.html>

Idaho Department of Environmental Quality. 2002. Idaho Waste Management Guidelines for Aquaculture Operations. [Online]. <http://www.deq.idaho.gov/water/prog_issues/waste_water/pollutant_trading/aquaculture_guidelines.pdf>

John, David E., and Joan B. Rose. 2005. Review of Factors Affecting Microbial Survival in Groundwater. *Environ. Sci. & Technol.*, 39(19):7345-7356.

Kalita, P., and J. Harner. 1998. In *Proc. of the International Conference on Engineering in Agriculture*. Perth, Australia.

Kansas State University (Department of Agronomy and Department of Agricultural Economics). 2003. Water Quality Best Management Practices, Effectiveness, and Cost for Reducing Contaminant Losses from Cropland. Kansas State University Agricultural Experiment Station and Cooperative Extension Service. Manhattan, Kan. Available: <http://www.oznet.ksu.edu/library/h20ql2/mf2572.pdf>

Kaplan, L.A., L.J. Standley, J.D. Newbold, J.H. Standridge, A.L. Mager, S.M. Kluender, L.L. Peterson, D.B. Smith, W.C. Hession, and P. Luitweiler. 2002. *Evaluation of Sources of Pathogens and NOM in Watersheds*. Denver, Colo.: AwwaRF and AWWA.

Keaton, M.R., P.K. Kalita, J.P. Harner, and S.K. Starrett. 1998. Evaluation of Vegetative Filter Strips for Nutrient Removal from Feedlot Runoff. ASAE International Annual Meeting, Orlando, FL. ASAE Paper No. 98-2014, St. Joseph, Mich.: ASAE.

Kellogg, R.L., C.H. Lander, D.C. Moffitt, and N. Gollehon. 2000. Manure Nutrients Relative to the Capacity of Cropland and Pastureland to Assimilate Nutrients. Publication NPS 00-0569. USDA-NRCS Economic Research Service, Washington, D.C.

Kendall, Carol. 2004. Tracing Sources of Agricultural N Using Isotopic Techniques. In *USEPA Regional Science Workshop: Animal Feeding Operations (AFOs)* - Science and Technical Support Needs. College Park, Md. Available: <http://www.epa.gov/osp/regions/afo_rep.pdf>

Ketelaars, H., G. Medema, L. van Breemen, D. Van der Kooij, P. Nobel, and P. Nuhn. 1995. Occurrence of *Cryptosporidium* Oocysts and *Giardia* Cysts in the River Meuse and Removal in the Biesbosch Reservoirs. *Aqua*. 108-111.

Kim, W.K., and P.H. Patterson. 2003. Effects of Minerals on Activity of Microbial Uricase to Reduce Ammonia Volatilization in Poultry Manure. *Poultry Science*. 82:223-231.

Klickitat County (Washington). 2005. Compost and Organics [Online]. Klickitat County, Goldendale, WA. Available: <http://www.klickitatcounty.org/SolidWaste/default.asp?fCategoryIDSelected=965105457> [cited July 18, 2006]

Kolodziej, E.P., T. Harter, and D.L. Sedlak. 2004. Dairy Wastewater, Aquaculture and Spawning Fish as Sources of Steroid Hormones in the Aquatic Environment. *Environ. Sci. & Technol.*, 38(23):6377-6384.

Kolpin, Dana W., Edward T. Furlong, Michael T. Meyer, E. Michael Thurman, Steven D. Zaugg, Larry B. Barber, and Herbert T. Buxton. 2002. Pharmaceuticals, Hormones, and Other Organic Wastewater Contaminants in U.S. Streams, 1999-2000: A National Reconnaissance. *Environ. Sci. & Technol.*, 36(6):1202-1211. Available : <http://pubs.acs.org/hotartcl/est/es011055j_rev.html>

LeChevallier, M., and K. Au. 2001. *Impact of Treatment on Microbial Water Quality: A Review Document on Treatment Efficiency to Remove Pathogens.* Washington, D.C.: World Health Organization.

LeChevallier, M.W., W.D. Norton, M. Abbaszadegan, and T.M. Atherholt. 1998. *Variation in Giardia and Cryptosporidium Levels in the Delaware River.* Voorhees, N.J.: American Water Works Service Company, Inc.

Lee, J.Y. and R.A. Deininger. 1999. A Rapid Method for Detecting Bacteria in Drinking Water. *Journal of Rapid Methods and Automation in Microbiology*, 7:135–145.

Lindsey, Bruce D., and Michele L. Koch. 2004. Determining Sources of Water and Contaminants to Wells in a Carbonate Aquifer Near Martinsburg, Blair County, Pennsylvania, By Use of Geochemical Indicators, Analysis of Anthropogenic Contaminants, and Simulation of Ground-Water Flow. USGS Scientific Investigations Report 2004-5124. Available: <http://pubs.usgs.gov/sir/2004/5124/pdf/sir2004-5124.pdf>

Long, Sharon C., Erica J. Mahar, Ruoting Pei, Catalina Arango, Erin Shafer, and Theodore H. Shoenberg. 2002. *Development of Source Specific Indicator Organisms for Drinking Water.* Denver, Colo.: AwwaRF.

Mallin, M.A., and L.B. Cahoon. 2003. Industrialized Animal Production - A Major Source of Nutrient and Microbial Pollution to Aquatic Ecosystems. *Population and Environment,* 24(5):369-385.

Mawdsley, J.L., A.E. Brooks, and R.J. Merry. 1996a. Movement of the Protozoan Pathogen Cryptosporidium Parvum Through Three Contrasting Soil Types. *Biol. Fert. Soils*, 21:30-36.

Mawdsley, J.L., A.E. Brooks, R.J. Merry and B.F. Pain. 1996b. Use of a Novel Soil Tilting Table Apparatus to Demonstrate Horizontal and Vertical Movement of the Protozoan Pathogen Cryptosporidium Parvum. *Biol. Fert. Soils*, 23:215-220.

McKinley, Blair, Malcolm Broome, and Larry Oldham. 2000. Poultry Nutrient Management Through Livestock Feedstuffs. Mississippi State, Miss.: Mississippi State University Cooperative Extension. Available: <http://msucares.com/pubs/misc/m1146.htm> (last modified September 20, 2002).

Medema, G., F. Schets, P. Teunis, and A. Havelaar. 1998. Sedimentation of Free and Attached *Cryptosporidium* Oocysts and *Giardia* Cysts in Water. *Appl. Environ. Microbiol*, 64:4460-4466.

Migliore L., G. Brambilla, A. Grassitellis and G. Dojmi di Delupis. 1993. Toxicity and Bio Accumulation of Sulfaphadimethoxine in Artmia (Crustacea Anostraca). *Int. J. Salt Lake Res.* 2:141-152.

Miller, Brian K., Brian J. MacGowan, and Richard P. Reaves. 2003. Are Constructed Wetlands a Viable Option for Your Waste Management System? West LaFayette, Ind.: Purdue University Cooperative Extension Service. Publication FNR-202-W. Available: <http://www.ces.purdue.edu/ extmedia/FNR/FNR-202.pdf>

Miller, Dan, and Ken Semmens. 2002. Waste Management in Aquaculture. Aquaculture Information Series. #AQ02-1. Morgantown, W.Va. : West Virginia University. Available: <http://www.wvu.edu/~agexten/aquaculture/waste02.pdf>

Minnesota Department of Health. 2003. City of Mankato Source Water Assessment. [Online]. <http://156.98.150.16/swa/surfwaterFile/1070009.pdf>

Moore, Keara B., Brenda Ekwurzel, Bradley K. Esser, G. Bryant Hudson, and Jean E. Moran. 2006. Sources of Groundwater Nitrate Revealed Using Residence Time and Isotope Methods. *Applied Geochemistry*, 21(6)1016-1029.

Moser, Mark A. 2006. Anaerobic Digesters Control Odors, Reduce Pathogens, Improve Nutrient Manageability, Can be Cost Competitive with Lagoons, and Provide Energy Too! USEPA [Online]. <http://www.epa.gov/agstar/resources/man_man.html>

Mullins, Gregory. 2001. Phosphorus, Agriculture & The Environment. Virginia Cooperative Extension, Publication Number 424-029, revised 2001. Available: <http://www.ext.vt.edu/pubs/grains/424-029/424-029.html>. [cited July 18, 2006]

Nahm, K. 2003. *Avian and Poultry Biology Reviews.* 14:151.

NDESC (National Dairy Environmental Stewardship Council). 2005. Cost-effective and Environmentally Beneficial Dairy Manure Management Practices. Sustainable Conservation. San Francisco, Calif. Available: <http://www.suscon.org/dairies/ndesc.asp or http://www.suscon.org/dairies/pdfs/COST%20EFFECTIVE%20AND%20ENVIRONMENTALLY.pdf>

NEIWPCC (New England Interstate Water Pollution Control Commission). 2000. Source Protection: A National Guidance Manual for Surface Water Supplies. Lowell, Mass.: NEIWPCC. Available: <http://www.neiwpcc.org> (click on "Publications", then click "Technical Guides" and find the report).

North, J., D. Schnoebelen, S. Grapp, and R. Hesemann. 2003. Evolution from a Conventional Well Field to a Riverbank-Filtration System. *Second International Riverbank Filtration Conference*. National Water Research Institute. Fountain Valley, Calif.

NRAES (Natural Resource, Agriculture, and Engineering Service). 1999. Field Guide to On-Farm Composting. Ithaca, NY: NRAES. NRAES-114. Available: <http://www.nraes.org/publications/nraes114.html>

NRAES (Natural Resource, Agriculture, and Engineering Service). 1992. On-Farm Composting Handbook. Ithaca, NY: NRAES. NRAES-54. Available: <www.nraes.org/publications/nraes54.html>

NRC (National Research Council, Committee on Animal Nutrition). 1989. *Nutrient Requirements of Horses*. Washington, D.C.: National Academy Press.

NSAIS (National Sustainable Agriculture Information Service). 2006. Farm-Scale Composting Resource List. NSAIS, Fayetteville, AR. Available: <www.attra.org/attra-pub/farmcompost.html>

NYSDEC (New York State Department of Environmental Conservation). 1996. *Agricultural Management Practices Catalogue for Nonpoint Source Pollution Prevention and Water Quality Protection in New York State*. Albany, N.Y.: NYSDEC, Bureau of Water Quality Management, Division of Water.

OCC (Oklahoma Conservation Commission). 2004. *Watershed Based Plan: Lake Eucha Watershed (Draft 3)*. Oklahoma City, Okla.: Oklahoma Conservation Commission.

ODEQ (Oklahoma Department of Environmental Quality). 2003. *Water Quality Modeling Analysis in Support of TMDL Development for Lake Eucha and Spavinaw Creek Watershed in Oklahoma - Draft Report.* Oklahoma Department of Environmental Quality. Oklahoma City, Okla.

Oskam, G. 1995. Main Principles of Water Quality Improvement in Reservoirs. *Aqua,* 44:23-29.

OWRB (Oklahoma Water Resources Board). 2001. *Draft - Water Quality Evaluation of the Eucha/Spavinaw Lake System*. Oklahoma City, Okla.: Oklahoma Water Resources Board.

OWRB (Oklahoma Water Resources Board). 2002. *Water Quality Evaluation of the Eucha/Spavinaw Lake System*. Oklahoma City, Okla.: Oklahoma Water Resources Board. Available: <http://www.owrb.state.ok.us/studies/reports/eucha-spav/eucha-spav.php>

Philadelphia Water Department. 2005. Draft Version: The Schuylkill River Watershed Source Water Protection Plan (Belmont & Queen Lane Surface Water Intakes). Philadelphia, Pa.: Philadelphia Water Department.

Philadelphia Water Department. 2002a. – Belmont & Queen Lane Treatment Plants (PWSID #1510001) Source Water Assessment Report. Philadelphia, Pa.: Philadelphia Water Department.

Philadelphia Water Department. 2002b. – Baxter Water Treatment Plant (PWSID #1510001) Source Water Assessment Report. Philadelphia, Pa.: Philadelphia Water Department.

Pollack, A.J., A.S.C. Chen, R.C. Haught, and J.A. Goodrich. 1991. *Options for Remote Monitoring and Control of Small Drinking Water Facilities*. Columbus, OH: Battelle Press.

Prescott, Annette, and Colin Fricker. 2002. Feasibility of Fast Response Testing for Coliform Bacteria in Distribution Systems. Denver, Colo.: AwwaRF.

PSU (Pennsylvania State University). 1992. Agricultural BMP Effectiveness Database. Pennsylvania State University, Department of Agricultural and Biological Engineering. Available: <http://iaspub.epa.gov/edr/edr_proc_qry.navigate?P_LIST_OPTION_CD=CSDIS&P_REG_AUTH_IDENTIFIER=1&P_DATA_IDENTIFIER=90637&P_VERSION=1>

Pyke, Grantley W., William C. Becker, Richard Head, and Charles R. O'Melia. 2003. *Impacts of Major Point and Non-Point Sources on Raw Water Treatability*. Denver, Colo.: AwwaRF.

Raucher, Robert S., and James Goldstein. 2001. *Guidance to Utilities on Building Alliances with Watershed Stakeholders*. Denver, Colo.: AwwaRF.

Robbins, R.W., J.L. Glicker, D.M. Bloem, and B.M. Niss. 1991. *Effective Watershed Management for Surface Water Supplies*. Denver, Colo.: AwwaRF.

Rose, J.B. and D.J. Grimes. 2001. "Reevaluation of Microbial Water Quality: Powerful New Tools for Detection and Risk Assessment." 18 pp. Washington, D.C.: American Academy of Microbiology.

Rosen, Barry H., R. Croft, E.R. Atwill, S. Wade, and S. Stehman. 2001. Waterborne Pathogens in Agricultural Watersheds. Natural Resource, Agriculture, and Engineering Service (NRAES) Cooperative Extension. Ithaca, N.Y. Available: <http://www.nraes.org/publications/nraes147.html>

Schneider, J. 1994. *Quimica Analitica*, 13:S34.

Schnoebelen, D., M. Turco, and J. North. 2003. Groundwater Flow and Water Quality – A Flowpath Study in the Seminole Well Field, Cedar Rapids, Iowa. In *Second International Riverbank Filtration Conference*. Fountain Valley, Calif.: National Water Research Institute.

Schulmeyer, P., and D. Schnoebelen. 1998. *Hydrology and Water Quality in the Cedar Rapids Area, Iowa, 1992-1996*. USGS Water-Resources Investigation Report 97-4261. Washington, D.C.: USGS.

Scott, Troy M., Joan B. Rose, Tracie M. Jenkins, Samuel R. Farrah, and Jerzy Lukasik. 2002. Microbial Source Tracking: Current Methodology and Future Directions. *Applied and Environmental Microbiology*, 68(12):5796-5803. Available: <http://aem.asm.org/cgi/content/full/68/12/5796>

Sharpley, A.N., T. C. Daniel, J.T. Sims, J. Lemunyon, R.A. Steven, and R. Parry. *Agricultural Phosphorus and Eutrophication*. US Department of Agriculture, Agricultural Research Service, ARS-149. Washington, D.C.: U. S. Govt. Printing Office.

Shore, L.S., D.L. Correll, and P.K. Chakraborty. 1995. In *Animal Waste and the Land-Water Interface*. Edited by K. Steele. Boca Raton, Fla.: Lewis Publications.

Simeral, Kenneth D. 1998. Using Constructed Wetlands for Removing Contaminants from Livestock Wastewater. Publication A-5-98. Columbus, Ohio: Ohio State University Extension. Available: <http://ohioline.osu.edu/a-fact/0005.html>

Smith, Robert J. 2005. State Doubles its Incentive to Haul Poultry Litter Out of Watersheds. *Arkansas Democrat-Gazette*. Little Rock, Ark. Available: <http://biosystems2.okstate.edu/poultry/upload/State_Doubles_Incentive.pdf>

SRAC (Southern Regional Aquaculture Center). 1999. *Characterization and Management of Effluents from Aquaculture Ponds in the Southeastern United States*. SRAC Publication No. 470. Available: <http://srac.tamu.edu/fulllist.cfm or http://srac.tamu.edu/tmppdfs/12877650-470fs.pdf>

Stone, K.C., P.G. Hunt, A.A. Szogi, F. J. Humenik, and J.M. Rice. 2000. Constructed Wetland Design and Performance for Swine Lagoon Waste Water Treatment. Presented at the 2000 ASAE Annual International Meeting, Milwaukee, Wis., July 9-12, 2000. Available: <http://www.florence.ars.usda.gov/publist/pdf/Man603.pdf>

Storm, Daniel E., and Michael J. White. 2003. Lake Eucha Basin SWAT 2000 Model Simulations Using New Row Crop/Small Grains Soil Test Data. Oklahoma State University, Biosystems and Agricultural Engineering Department. Stillwater, Okla. Available: <http://biosystems.okstate.edu/home/dstorm/eucha/modeling/new%20rowcrop%20stp%20runs%20-%207.pdf>

Sturdevant Rees, P.L., S.C. Long, R. Baker, D.H. Bordeau, R. Pei, and P.K. Barten. [N.d.] Development of Event-Based Pathogen Monitoring Strategies for Watersheds. *AwwaRF*. Forthcoming.

Telega, Stanley (Lee), and Peg Cook. 2003. Fact Sheet #5: What if My Operation is an AFO But Not a CAFO? Livestock and Poultry Environmental Stewardship (LPES) Curriculum, MidWest Plan Service. Ames, Iowa: Iowa State University. Available: <http://www.lpes.org/cafo/05FS_AFO.pdf>

Tetra Tech, Inc. 2003. *Getting in Step: A Guide for Conducting Watershed Outreach Campaigns*. EPA 841-B-03-002. Washington, D.C.: USEPA.

Tetra Tech, Inc. 2002. State Compendium: Programs and Regulatory Activities Related to Animal Feeding Operations. Washington, D.C.: USEPA, Office of Wastewater Management. Available: <http://www.epa.gov/npdes/pubs/statecom.pdf>

Texas Commission on Environmental Quality. 2003. Source Water Assessment Report: Prepared for PWS #1880001 Amarillo Municipal Water System.

Trask, J.R., P.K. Kalita, M.S. Kuhlenschmidt, R.D. Smith, and T.L. Funk. 2004. Overland and Near-Surface Transport of *Cryptosporidium parvum* from Vegetated and Nonvegetated Surfaces. *Jour. Environ. Qual.,* 33:984-993. Available: <http://jeq.scijournals.org/cgi/content/full/33/3/984>

Tyrrel, S.F, and J.N. Quinton. 2003. Overland Flow Transport of Pathogens From Agricultural Land Receiving Faecal Wastes. *Jour. Appl. Microbiol.*, 94:87S-93S.

Tyson, Ted W. 1996. *Constructed Wetlands for Animal Waste Treatment.* Publication ANR-965. Alabama Cooperative Extension System. Available: <http://www.aces.edu/pubs/docs/A/ANR-0965>

University of California (Division of Agriculture and Natural Resources, Committee of Experts on Dairy Manure Management). 2005. Managing Dairy Manure in the Central Valley of California (June 1995 Revision). Available: <http://www.waterboards.ca.gov/centralvalley/available_documents/dairies/uc-committee-of-experts-rpt.pdf>

USDA (United States Department of Agriculture). 1998. 1998 Census of Aquaculture. USDA, National Agricultural Statistics Service. [Online]. <http://www.nass.usda.gov/census/census97/aquaculture/aquaculture.htm>. Data obtained from Table 1 (<http://www.nass.usda.gov/census/census97/aquaculture/table1.pdf>) [cited May 1, 2006]

USDA (United States Department of Agriculture). 2004. 2002 Census of Agriculture. USDA, National Agricultural Statistics Service. Available: <http://www.nass.usda.gov/census/census02/volume1/us/USVolume104.pdf>; on-line searchable database available:<http://agcensus.mannlib.cornell.edu> or <http://www.nass.usda.gov/Census_of_Agriculture>

USDA (United States Department of Agriculture). 1999. Agricultural Waste Management Field Handbook (National Engineering Handbook, Part 651). USDA Natural Resources Conservation Service. 210-VI, NEH-651. First published in April 1992; last updated June 1, 1999. Available: <http://www.wcc.nrcs.usda.gov/awm/awmfh.html>

USDA/USEPA (United States Department of Agriculture/United States Environmental Protection Agency). 1999. USDA/USEPA Unified National Strategy for Animal Feeding Operations. Available: <http://cfpub.epa.gov/npdes/afo/ustrategy.cfm>

USEPA (United States Environmental Protection Agency). 2005a. *Case Studies of Local Source Water Protection Programs*. Office of Ground Water and Drinking Water. Washington, D.C.: USEPA. Available: <http://www.epa.gov/safewater/protect/casesty/index.html>

USEPA (United States Environmental Protection Agency). 2005b. Microbial Source Tracking Guide Document. Office of Research and Development. EPA/600-R-05-064. Cincinnati, Ohio: USEPA. Available: <http://www.ces.purdue.edu/waterquality/resources/MSTGuide.pdf> or <http://www.calcoast.org/news/MSTGuide.pdf>

USEPA (United States Environmental Protection Agency). 2005c. *Financial Assistance Summaries for AFOs*. Office of Water. Washington, D.C.: USEPA. Available: <http://www.epa.gov/npdes/pubs/financial_assistance_summaries.pdf>

USEPA (United States Environmental Protection Agency). 2005d. Draft - Handbook for Developing Watershed Plans to Restore and Protect our Waters. Office of Water. EPA 841-B-05-005. Washington, D.C.: USEPA. Available: <http://www.epa.gov/owow/nps/watershed_handbook>

USEPA (United States Environmental Protection Agency). 2005e. Section 319 Nonpoint Success Stories Website. Office of Water. EPA 841-F-05-004. Washington, D.C.: USEPA. Available: <www.epa.gov/owow/nps/Success319>

USEPA (United States Environmental Protection Agency). 2004a. Managing Manure Nutrients at Concentrated Animal Feeding Operations. Office of Water. EPA-821-B-04-009. Washington, D.C.: USEPA. Available: <http://www.epa.gov/npdes/pubs/cafo_manure_guidance.pdf>

USEPA (United States Environmental Protection Agency). 2004b. Risk Assessment Evaluation for Concentrated Animal Feeding Operations. Office of Research and Development National Risk Management Research Laboratory. EPA/600/R-04/042. Cincinnati, Ohio: USEPA. Available: <http://www.epa.gov/ORD/NRMRL/pubs/600r04042/600r04042.pdf>

USEPA (United States Environmental Protection Agency). 2004c. Effluent Limitations Guidelines and New Source Performance Standards for the Concentrated Aquatic Animal Production Point Source Category. *Fed. Reg.* 69(162):51891-51930

USEPA (United States Environmental Protection Agency). 2003a. National Pollutant Discharge Elimination System Permit Regulation and Effluent Limitation Guidelines and Standards for Concentrated Animal Feeding Operations (CAFOs); Final Rule. *Fed. Reg.* (40 CFR Parts 9, 122, 123, and 412), 68(29):7175-7274. Available: <http://cfpub.epa.gov/npdes/afo/cafofinalrule.cfm>

USEPA (United States Environmental Protection Agency). 2003b. NPDES Permit Writer's Guidance Manual and Example NPDES Permit for Concentrated Animal Feeding Operations. EPA-833-B-04-001. Washington, D.C.: USEPA. Available: http://cfpub.epa.gov/npdes/afo/info.cfm.

USEPA (United States Environmental Protection Agency). 2003c. Producers' Compliance Guide for CAFOs. EPA 821-R-03-010. Washington, D.C.: USEPA. Availabe: <http://cfpub.epa.gov/npdes/afo/ compliance.cfm>

USEPA (United States Environmental Protection Agency). 2003d. National Management Measures for the Control of Nonpoint Pollution from Agriculture. Office of Water. EPA 841-B-03-004. Washington, D.C.: USEPA. Available: <http://www.epa.gov/nps/agmm>

USEPA (United States Environmental Protection Agency). 2003e. *Introduction to EPA's Drinking Water Source Protection Programs.* Office of Ground Water and Drinking Water. Washington, D.C.: USEPA. Available: <http://www.epa.gov/OGWDW/dwa/electronic/presentations/swp/swp.ppt#1082>

USEPA (United States Environmental Protection Agency). 2003f. Annotated Bibliography of Source Water Materials. Office of Ground Water and Drinking Water. EPA 816-F-03-010. Washington, D.C.: USEPA. Available: <http://www.epa.gov/safewater/protect/swpbibliography>

USEPA (United States Environmental Protection Agency). 2002a. Environmental and Economic Benefit Analysis of Final Revisions to the National Pollutant Discharge Elimination System Regulation and the Effluent Guidelines for Concentrated Animal Feeding Operations. EPA-821-R-03-003. Washington, D.C.: USEPA. Available (in four parts - scroll down): <http://cfpub.epa.gov/npdes/afo/cafodocs.cfm>

USEPA (United States Environmental Protection Agency). 2002b. Alternative Technologies/Uses for Manure (Draft Report). Office of Wastewater Management. EPA 833-D-02-001. Washington, D.C.: USEPA. Available: <http://www.epa.gov/npdes/pubs/cafo_report.pdf>

USEPA (United States Environmental Protection Agency). 2002c. Development Document for the Proposed Revisions to the National Pollutant Discharge Elimination System (NPDES) Regulation and the Effluent Guidelines for Concentrated Animal Feeding Operations. Office of Water. EPA-821-R-03-001. Washington D.C.: USEPA. Available: <http://cfpub.epa.gov/npdes/afo/cafodocs.cfm>

USEPA (United States Environmental Protection Agency). 2002d. *Source Water Protection: Best Management Practices and Other Measures for Protecting Drinking Water Supplies.* Office of Ground Water and Drinking Water. Washington, D.C.: USEPA. Available: <http://www.epa.gov/watertrain/pdf/swpbmp.pdf>

USEPA (United States Environmental Protection Agency). 2001a. Source Water Protection Practices Bulletin: Managing Livestock, Poultry, and Horse Waste to Prevent Contamination of Drinking Water. Office of Water. EPA-916-F-01-026. Washington, D.C.: USEPA. Available: <http://www.epa.gov/safewater/protect/pdfs/livestock.pdf>

USEPA (United States Environmental Protection Agency). 2001b. Source Water Protection Practices Bulletin: Managing Agricultural Fertilizer Application to Prevent Contamination of Drinking Water. Office of Water. EPA-916-F-01-028. Washington, D.C.: USEPA. Available: <http://www.epa.gov/safewater/protect/pdfs/fertilizer.pdf>

USEPA (United States Environmental Protection Agency). 2001c. Protecting and Restoring America's Watersheds: Status, Trends, and Initiatives in Watershed Management. Office of Water. EPA 840-R-00-001. Washington, D.C.: USEPA. Available: <http://www.epa.gov/owow/protecting>

USEPA (United States Environmental Protection Agency). 2000. *Watershed Success Stories – Applying the Principles and Spirit of the Clean Water Action Plan.* Washington, D.C.: USEPA. Available: <http://water.usgs.gov/owq/cleanwater/success/index.html>

USEPA (United States Environmental Protection Agency). 1999. Protecting Sources of Drinking Water: Selected Case Studies in Watershed Management. Office of Water. EPA 816-R-98-019. Washington D.C.: USEPA. Available: <http://www.epa.gov/safewater/swp/swpcases.pdf>

USEPA (United States Environmental Protection Agency). 1997a. Constructed Wetlands for Animal Waste Treatment: A Manual on Performance Design and Operation with Case Histories. EPA-855B97001. Washington, D.C.: USEPA. Available: <http://yosemite.epa.gov/water/owrccatalog.nsf/0/caf5289c155133ef85256d83004fd7e0?OpenDocument>

USEPA (United States Environmental Protection Agency). 1997b. Constructed Wetlands And Wastewater Management For Confined Animal Feeding Operations. EPA Number: 855K97001. Washington, D.C.: USEPA. Available: <http://yosemite.epa.gov/water/owrccatalog.nsf/0/676abe7d693e905585256d83004fd9c1?OpenDocument>

USEPA (United States Environmental Protection Agency). 1997c. State Source Water Assessment and Protection Programs. Final Guidance. Office of Water. EPA 816-R-97-009. Washington D.C.: USEPA. Available: <http://www.epa.gov/ogwdw/swp/swp.pdf>

Van Breemen, L., H. Ketelaars, W. Hoogenboezem, and G. Medema. 1998. Storage Reservoirs-A First Barrier For Pathogenic Micro-Organisms In The Netherlands. *Water Science and Technology*, 37(2):253-260.

Van Herk, F.H., T.A. McAllister, C.L. Cockwill, N. Guselle, F.J. Larney, J.J. Miller, and M.E. Olson. 2004. Inactivation of *Giardia* Cysts and *Cryptosporidium* Oocysts in Beef Feedlot Manure. *Compost Science & Utilization*, 12(3):235-241.

Venter, S.N. 2000. "Rapid microbiological Monitoring Methods: The Status Quo." *The Blue Pages*. London, UK: International Water Association.

Vogel, J.R., S.I. Harris, T.B. Coplen, E.W. Rice, and I.M. Verstraeten. 2005a. Microbe Concentrations, Laser Particle Counts, and Stable Hydrogen and Oxygen Isotope Ratios in Samples from a Riverbank Filtration Study, Platte River, Nebraska, 2002 to 2004. US Geological Survey. Data Series 133. Reston, Va.: USGS. Available: <http://pubs.usgs.gov/ds/2005/133>

Vogel, J.R., I.M. Verstraeten, T.B. Coplen, E.T. Furlong, M.T. Meyer, and L.B. Barber. 2005b. Occurrence of Selected Pharmaceutical and Non-Pharmaceutical Compounds and Stable Hydrogen and Oxygen Isotope Ratios in a Riverbank Filtration Study, Platte River, Nebraska, 2001 to 2003, Volume 1. US Geological Survey. Data Series 117. Reston, Va.: USGS. Available: <http://pubs.usgs.gov/ds/2005/117>

Wagner, K., and S. Woodruff. 1997. Phase I Clean Lakes Project: *Diagnostic and Feasibility Study of Lake Eucha*. Oklahoma City, Okla.: Oklahoma Conservation Commission.

Walker, Forbes. 2000. Best Management Practices for Phosphorus in the Environment. PB 1645-5M-6/00. University of Tennessee Agricultural Extension Service. Available: <http://www.utextension.utk.edu/publications/pbfiles/PB1645%20.pdf>

Walker, Mark, Katherine Leddy, and Elaine Hagar. 2001. Effects of Combined Water Potential and Temperature Stresses on *Cryptosporidium parvum* Oocysts. *Applied and Environmental Microbiology,* 67(12):5526–5529. Available: <http://www.pubmedcentral.nih.gov/articlerender.fcgi?artid=93339#B6>

Wang, J., S. Hubbs, and R. Song. 2002. *Evaluation of Riverbank Filtration as a Drinking Water Treatment Process*. Denver, Colo.: AWWA and AwwaRF

Williamson, T. S., P. K. Kalita, J.P. Harner, G.L. Marchin, and C.W. Rice. 1999. Coliform Reduction by Filter Strips in Kansas Watershed. ASAE Annual International Meeting, Toronto, Canada. ASAE Paper No. 99-2070, St. Joseph, Mich.: ASAE.

Wimberly, J., and H.L. Goodwin. 2000. "Alternative Poultry Litter Management in the Eucha/Spavinaw Watershed." Report to the Tulsa Metropolitan Utility Authority, Tulsa, Okla.

Wing, Steve, Stephanie Freedman, and Lawrence Band. 2002. The Potential Impact of Flooding on Confined Animal Feeding Operations in Eastern North Carolina. *Environmental Health Perspectives*, 110(4). Available: <http://www.ehponline.org/docs/2002/110p387-391wing/abstract.html>

Wright, Peter, and Curt Gooch. 2005. Water Quality Issues from Dairy Farms in the Northeast. Presented at the Workshop on Concentrated Animal Feeding Operations (CAFOs): Water Utility Issues and Regulatory Controls (AwwaRF Project #3020 meeting), Cleveland, Ohio, September 19-20, 2005.

Yagow, G., T. Dillaha, J. Pease, D. Kibler, and D. Bosch. 2002. A BMP Database for Nutrient Reduction. In *Proc. of the March 11-13, 2002 Conference - Total Maximum Daily Load (TMDL) Environmental Regulations*. Fort Worth, Texas. St. Joseph, Mich.: ASABE. 250-255. Available: <http://asae.frymulti.com/abstract.asp?aid=7565&t=2>

Zhang, Q., K.R. Mankin, and L.E. Erickson. 2001. Modeling Transport Of *Cryptosporidium Parvum* Oocysts In Overland Flow. In *Proc. of the 2001 Conference on Environmental Research*. Available: <http://www.engg.ksu.edu/HSRC/01Proceed/docs/32.pdf>

ABBREVIATIONS

ADU	animal disposal unit
AFO	animal feeding operation
AMA	Agricultural Management Assistance (program)
APHA	American Public Health Association
AquaNIC	Aquaculture Network Information Center
ASAE	American Society of Agricultural Engineers
ASABE	American Society of Agricultural Biological Engineers (formerly ASAE)
ASCE	American Society of Civil Engineers
AWWA	American Water Works Association
AwwaRF	Awwa Research Foundation
BGH	bovine growth hormone
BMP	best management practice
BOD	biochemical oxygen demand
BST	Bovine Somatotropin
^{o}C	degrees Celcius
CAFO	concentrated animal feeding operation
CFR	Code of Federal Regulations
cfu	colony forming units
CNMP	comprehensive nutrient management plan
cm	centimeter
COD	chemical oxygen demand
CRC	Cooperative Research Centre for Water Quality and Treatment
CSREES	Cooperative State Research, Education, and Extension Service
d	day
d-a	per day per animal
DBP	disinfection byproduct
DBPR	Disinfectant Byproduct Rule
DEQ	Department of Environmental Quality
ECL	Environmental Conservation Law
f.a.	finished animal
ft^3	cubic foot
ELGs	effluent limitation guidelines and standards
EQIP	Environmental Quality Incentives Program
GIS	Geographic Information System
GWUDI	ground water under direct influence (of surface water)
h	head (of animals)
HPC	heterotrophic plate counts

INCOG	Indian Nations Council of Governments
kg	kilogram
L	liter
lbs	pounds
LPES	Livestock and Poultry Environmental Stewardship
LT2ESWTR	Long-Term 2 Enhanced Surface Water Treatment Rule
MCL	maximum contaminant level
mg	milligram
mL	milliliter
mmho	millimho
mS	milliSiemen
MST	microbial source tracking
μg	microgram
NAIC	North American Industry Code
NASS	National Agricultural Statistics Service (part of USDA)
NCDENR	North Carolina Department of Environment and Natural Resources
NDESC	National Dairy Environmental Stewardship Council
NMP	nutrient management plan
NPDES	National Pollutant Discharge Elimination System
NRCS	Natural Resources Conservation Service (part of USDA)
ntu	nephelometric turbidity unit
NYCRR	New York Code of Rules and Regulations
NYSDOH	New York State Department of Health
OCC	Oklahoma Conservation Commission
ORP	oxidation-reduction potential
OWRB	Oklahoma Water Resources Board
PAC	Project Advisory Committee
PADEP	Pennsylvania Department of Environmental Protection
PLAT	phosphorus loss assessment tool
ppm	part per million
PWD	Philadelphia Water Department
QAPP	quality assurance project plan
RBF	riverbank filtration
SAN	Schuylkill Action Network
SLWAP	Skaneateles Lake Watershed Agricultural Program
SPCC	Spill Prevention, Control, and Countermeasures
SRAC	Southern Regional Aquaculture Center
SWAP	Source Water Assessment Program

TCFA	Texas Cattle Feeders Association
TKN	total Kjeldahl nitrogen
TMDL	Total Maximum Daily Load
TOC	total organic carbon
TS	total solids
USDA	United States Department of Agriculture
USEPA	United States Environmental Protection Agency
USGS	United States Geological Survey
UV	ultraviolet light
VOC	volatile organic compound
VS	volatile solids
w.b.	wet basis
WFP	Whole Farm Plan
WRAS	Watershed Restoration Action Strategy
WTR	water treatment residual

IWA Publishing's authorised EU representative for General Product Safety Regulations is Diane D'Arras, 15 rue Duret, 75116 Paris, France, e-mail: safety@iwap.co.uk.

Printed and bound by CPI Group (UK) Ltd, Croydon, CR0 4YY
18/09/2026
02218938-0004